T0136233

Atlas of Coastal Ecosystems
in the Western Gulf of California

Isla Monserrat, located southeast of Loreto in the Gulf of California showing patches of Pliocene limestone (beige) on Miocene volcanic rocks (rust red). Photo credit: David H. Backus

Atlas of Coastal Ecosystems in the Western Gulf of California

TRACKING LIMESTONE DEPOSITS ON THE MARGIN OF A YOUNG SEA

EDITED BY Markes E. Johnson and Jorge Ledesma-Vázquez

THE UNIVERSITY OF ARIZONA PRESS | TUCSON

The University of Arizona Press
© 2009 The Arizona Board of Regents

www.uapress.arizona.edu

Library of Congress Cataloging-in-Publication Data
Atlas of coastal ecosystems in the western Gulf of California :
tracking limestone deposits on the margin of a young sea /
edited by Markes E. Johnson and Jorge Ledesma-Vázquez.
    p. cm.
Includes bibliographic references and index.
ISBN 978-0-8165-2530-0 (cloth : alk. paper) —
ISBN 978-0-8165-2229-3 (pbk. : alk. paper)
   1. Rocks, Carbonate—Mexico—California, Gulf of. 2. Limestone—
Mexico—California, Gulf of. 3. Geology, Stratigraphic—Pliocene.
4. Geology, Stratigraphic—Pleistocene. 5. Geology—Mexico—
California, Gulf of. 6. Coastal ecology—Mexico—California, Gulf of.
7. California, Gulf of (Mexico) —Geography.
I. Johnson, Markes E. II. Ledesma-Vázquez, Jorge.
QE471.15.C3A866 2010
552'.5809722—dc22   2009013145

The text of this book is printed on acid-free, archival-quality
128gsm FSC Neo Matte paper in order to be environmentally
friendly. Manufactured in China.

14 13 12 11 10 09   6 5 4 3 2 1

CONTENTS

The central theme of this volume is limestone production over the western Gulf of California through geological time traced from the present back 5.5 million years. Chemically defined as calcium carbonate ($CaCO_3$), limestone is part of the carbon cycle, a process of vital importance to a world under impact by global warming. The Gulf of California is acknowledged as one of the richest bodies of water on the planet and is justly famous for its fish stocks and marine mammals. Typically undertaken by a diverse array of mollusks, corals, and coralline algae, limestone production comes from another part of the ecological web commonly found in coastal settings. This source has been largely overlooked in terms of carbon sequestration. The collected research found in these chapters provides insight into the scale of the process involving humble invertebrates and algae as played out in the Gulf of California. We hope it will be a starting point for more sophisticated studies along the same lines.

Plans for this volume on living and fossil components of coastal ecosystems in which carbonate production plays a major role were launched during the influential Gulf of California Conference held in Tucson, Arizona, on June 15–16, 2004. The editors, who have collaborated on geological and paleontological projects throughout the Baja California peninsula since 1990, saw an opportunity to expand and consolidate studies using satellite images as a guide to past and present coastal zones along the gulf shores of the Baja California peninsula from Cabo San Lucas in the south to the Colorado River delta in the north, including all major and many minor islands in the Gulf of California. Prior to 2004, much of our research was focused around Mulegé and Loreto in Baja California Sur. The false-color imagery of satellite photos brings landscapes to life with a vibrancy that approaches art. It also makes possible the easy detection of places

where limestone deposits are well exposed. Isla Monserrat, which features patches of Pliocene limestone preserved at unusually high elevations (frontispiece), is a good example of the hard-to-reach places we took care to explore during the ensuing years. Most of the chapters in this volume include satellite images that are as beautiful as they are rich in scientific content.

When puzzled by a geological field problem requiring an environmental interpretation, the standing joke between us called for seeking the solution outside the rocks in a modern setting usually within sight. "Just turn around and find the answer on the shore," we are prone to say when all else fails on the rock outcrop. Although not marine ecologists, our grounding in geology gives us a healthy respect for the living Earth, and the call for conservation has reached our ears. Biennial meetings of the Sociedad Geológica Peninsular, convened in Ensenada, Loreto, and La Paz during the last two decades, allowed us to meet with colleagues who do have credentials as marine ecologists, and we have solicited their involvement. Thus, the plan for this volume was to pair chapters on geological aspects of a given coastal ecosystem from the past with chapters on its modern counterpart. Fourteen colleagues joined us in this enterprise, bringing expertise beyond our capabilities. Some are leading authorities in their field, and others represent the upcoming generation of researchers with fresh perspectives (see the list of contributors with institutional affiliations at back of the volume).

Chapter 1 provides an introduction to the physical aspects of the Gulf of California with an emphasis on geological origins. The pre-Pliocene history of marine flooding is an important topic generally not covered in other works. The visual treat embedded in this chapter is a quartet of color satellite images showing summer/winter phytoplankton concentrations and summer/winter sea-surface temperatures. Robust environmental patterns captured by these images underscore the Gulf of California's enormous fecundity at the base of the food pyramid.

The physical and biological characteristics of rocky shores are discussed in chapter 2. Tables on physical geography provide new information on the breakdown of rock types spread over nearly 3,000 km of coastline. Extensive tables on biological geography provide fresh results combed from the historic 1940 collecting trip by Ed Ricketts and John Steinbeck. The related chapter 3 looks at rocky shores that developed during the Pliocene and Pleistocene, now fully exposed on land. The account provides previously unpublished perspectives on former rocky shores in remote places, such as Isla Angel de la Guarda and San Francisquito in northern Baja California.

Chapters 4 and 5 cover modern corals and coral reefs in the Gulf of California as contrasted with their fossil counterparts. Today coral reefs are restricted to the far southern part of the gulf, whereas the distribution of corals and reefs during the Pleistocene and Pliocene was much more extensive. The distribution of modern coralline red algae in the form of rhodoliths and their fossil equivalents are dealt with in chapters 6 and 7. Both today and in the past, the production and wide distribution of these calcareous, nonattached algae in the Gulf of California are astonishing. Chapter 8 examines modern clam (bivalve) populations affected by the Colorado River delta and their ecological demise since the 1930s, when reduced water flow resulted from the construction of Hoover Dam. Chapter 9 looks at fossil clam banks, originally from an entirely marine setting but now the major component of extensive limestone deposits on land. Part of this story features oysters that early in the twentieth century were still common in the gulf, but are now virtually extinct. A novel part of the same story involves the effect of Pliocene earthquakes on clam banks.

Chapters 10 and 11 focus on modern and ancient coastal sand dunes around the Gulf of California. The former represents the first treatment of its kind to quantify the amount of carbonate sand in coastal sand-dune systems on the gulf coast of the Baja California peninsula. Like chapter 2, this contribution uses satellite imagery as a fundamental reconnaissance tool. The related chapter 11 provides fresh information on extensive limestone deposits that represent former sand dunes. Finally, modern and past hydrothermal springs are reviewed in chapter 12. Although it may seem a peculiar topic with which to conclude

the book, the Gulf of California is an actively growing seaway, and the tectonic forces shaping it generate heat released through fault-related springs on land, in shallow coastal waters, and in much deeper offshore waters. Evidence suggests that the gulf's biological fabric has been and continues to be enhanced by hydrothermal activity, most notably with regard to vital nursery settings in shallow embayments.

A special feature of this volume is a set of 26 Advanced Spaceborne Thermal Emission and Reflection Radiometer (ASTER) images acquired by the authors and used for reconnaissance studies undertaken in preparation for writing of chapters 2, 3, 7, 9, 10, and 11. The set is also stored as an archive on CD (included with this volume), and individual scenes are keyed with reference to the separate geographic maps for the northern state of Baja California and southern state of Baja California Sur (see ASTER images in the back). Overlap of the 26 images provides nearly 100 percent coverage of the gulf shores of the Baja California peninsula and associated islands in the Gulf of California. In this sense, the volume constitutes a true atlas covering the region's coastal settings. We hope it will prove useful to a wide range of geologists, paleoecologists, ecologists, conservation biologists, and naturalists of all stripes interested in following other topics not discussed in this volume.

Publication of a research volume requires the support of many individuals and institutions. This project would have floundered without the early and steady support of Allyson Carter, editor-in-chief at the University of Arizona Press. David H. Backus, research scientist at Williams College, selected and processed all satellite images found in the text and the archive on disk. The facilities of the GIS and Remote Sensing Laboratory in the Schow Library of Williams College were indispensable in the execution of this technical work, as well as the reconnaissance studies that followed. Grants from the Bronfman Science Fund at Williams College were allocated for the purchase of the 26 ASTER images and provided summer stipend support for visiting faculty and students engaged in studies on remote sensing. Lisa DiDonato Brousseau, our copyeditor for this volume, applied her considerable skills in hunting down all the various inconsistencies and errors that creep into a project of this kind, and she showed great patience with us in doing so.

We are especially grateful to the donors of the Petroleum Research Fund (American Chemical Society) for a substantial publication subsidy to the University of Arizona Press for production of this volume. Support from the Petroleum Research Fund through five consecutive grants since 1990 allowed us to enhance our knowledge on topics covered in chapters 2, 3, 7, 9, 10, and 11. The U.S. National Science Foundation and its Mexican counterpart, Consejo Nacional de Ciencia y Tecnología (CONACYT), provided additional support for international cooperation. Williams College and the Universidad Autónoma de Baja California (Ensenada) subsidized field courses for our students during January 1990, 1992, 1994, 1997, 1999, and 2005 and likewise supported thesis projects by students who chose to continue research under our supervision. All larger islands in the Gulf of California belong to nature reserves that require research permits for access. We thank the director from Parque Nacional Bahía de Loreto and the directors from Centro Instituto Nacional de Anthropologia e Historia for Baja California Sur and the northern Baja California district around Bahía de Los Angeles for their full cooperation. A special debt of gratitude is owed Leon Fichman of Baja Outpost in Loreto, who saw to our logistical support in visiting Islas Coronados, Carmen, Monserrat, and San José.

With respect to chapter 1, Ledesma-Vázquez thanks NSF-CONACYT and the Universidad Autónoma de Baja California for support under project 52301.

Regarding chapter 4, Reyes-Bonilla and López-Pérez acknowledge the staff and administration of many protected areas (Reserva de la Biosfera Bahía de Los Angeles y Canales de Ballenas y Salsipuedes, Parques Nacionales Loreto y Cabo Pulmo, Áreas de Protección de Flora y Fauna Islas del Golfo de California y Cabo San Lucas) and financing agencies and nongovernmental organizations (CONACYT, Comisión Nacional para el Conocimiento y Uso de la Biodiversidad, Fondo Mexicano para la Conservación de la Naturaleza, Pronatura Noroeste, Sociedad de Historia Natural Niparajá, A. C., PADI Aware Foundation, Programa de Mejoramiento del Profesorado, SEP) for their support. They also are grateful to the many colleagues and students who collaborated with them in the field and in the analysis of data over the past decade.

Early drafts of chapter 5 were improved by comments made by H. Reyes-Bonilla, K. G. Johnson, and M. E. Johnson. In addition, J. Ledesma-Vázquez, G. Barba, M. E. Johnson, D. Backus, A. Morales, S. Scarry, D. Paz, T. Herrera, M. Mora, and B. López assisted with collecting and preparing fossil specimens. López-Pérez's research was supported by a Doctoral Fellowship from CONACYT, and he received additional support from Project AS007 CONABIO and SEMARNAT-CONACYT México, Universidad Autónoma de Baja California, Universidad Autónoma de Baja California Sur, University of Iowa, Geological Society of America, the Max and Lorraine Littlefield Fund, and shared travel funds from Williams College.

Regarding chapter 6, Steller thanks Lynn McMasters at the Moss Landing Marine Laboratories for drafting the map in figure 6.3.

Research for chapter 10 on modern coastal dunes was supported by stipends during the summer of 2006 to David H. Backus and Caroline S. Doctor from the Sperry Fund in the Department of Geosciences at Williams College. Doctor compiled the raw data derived from satellite images on the distribution of coastal dunes.

With respect to chapter 12 on hydrothermal springs, Forrest acknowledges funding from the David and Lucile Packard Foundation, the Earl and Ethel Meyers Oceanographic Trust Fund, the PADI Aware Foundation, and the Charles H. Stout Foundation.

Finally, Ledesma-Vázquez thanks NSF-CONACYT and Universidad Autónoma de Baja California under project *Atlas de Ecosistemas Costeros en el Golfo de California: Pasado y Presente*. Additional thanks go to D. Steller, M. Foster, C. Mitchell, K. Mitchell, A. Melwani, R. M. Prol-Ledesma, C. Canet, R. Price, T. Pichler, J. Kulongoski, C. True, D. Darling, and K. Rowell for useful comments on early drafts of the manuscript.

Markes E. Johnson and **Jorge Ledesma-Vázquez**
January 2009

Figure 1.1. Merger of NASA images captured by Aqua/Modis sensor on November 30, 2003, with seafloor information from a collection of surveys by DANA07RR, DANA08RR (SIO), and other bathymetric data. Localities and basins from north to south: SF, San Felipe; UDB, upper Delfín basin; BA, Bahía de Los Angeles; SO, San Francisquito; GB, Guaymas basin; BC, Bahía Concepción; CB, Carmen basin; EM, El Mangle; LO, Loreto; FB, Farallon basin; PB, Pescadero basin; AB, Alarcón basin. Image merger credit: Alejandro Hinojosa-Corona

1 Gulf of California Geography, Geological Origins, Oceanography, and Sedimentation Patterns

Jorge Ledesma-Vázquez, Markes E. Johnson, Oscar Gonzalez-Yajimovich, and Eduardo Santamaría-del-Angel

A combination of climate, regional winds, and marine circulation stimulates the replenishment of nutrients that reach the Gulf of California through upwelling of deep water ultimately from the Pacific Ocean. The marine food web expands like a pyramid from a host of single-celled plants and animals at the broad base to mammals such as the humpback whale, sea lion, and dolphin at the apex. Each tier within the web depends on the level below for its productivity. The more phytoplankton and zooplankton are able to reproduce, the more food is available for consumption at all higher levels. Not only the rich fisheries of the gulf depend on this chain, but also commercially harvested marine invertebrates such as crustaceans (shrimp) and pectens (scallops).

Many other marine invertebrates collectively contribute in a significant way to the geology of the gulf through a phenomenon known as the carbonate factory, where "carbonate" refers to calcium carbonate ($CaCO_3$). Molecules of calcium carbonate provide the basic building blocks for organisms such as corals, coralline red algae, and bivalves (clams). During the brief geological history of the Gulf of California (little more than 5.5 million years in most places), these organisms have lived, died, and left behind their durable carbonate skeletons in enormous quantities.

Prior to cementation of the skeletal parts that comprise solid limestone, the bits and pieces of calcium carbonate produced by many invertebrates and some algae accumulate as loose organic sediment on the shallow seabed, on beaches, and (under special circumstances) on coastal dunes. The Gulf of California and its islands constitute a laboratory where these deposits can be observed accumulating today under the influence of wind and waves and where the corresponding limestone byproducts of past geological epochs still stand as massive monuments to the region's natural history.

Intersection of Geography, Oceanography, and Climate

The Gulf of California (fig. 1.1) is a marginal sea located between the Mexican mainland and the Baja California peninsula. It is 1,100 km long and 180 km wide at its mouth (Argote et al. 1995), with an enclosed area that approaches 210,000 km^2 (Roden and Groves 1959). There are 37 named islands with a cumulative area of 2,850 km^2 (Carreño and Helenes 2002). One of the smallest, Isla Patos (0.45 km^2), is located near Isla Tiburón, which is the largest (1,223.5 km^2). Except for Tiburón, which sits near the Mexican mainland, all the larger islands are closely associated with the Baja California peninsula. The physical geography of the peninsula's gulf coast and related islands has a composite shoreline approaching 3,000 km in length, as described in chapter 2. Pliocene-Pleistocene counterparts are covered in chapter 3.

Together with Isla Tiburón, Isla Angel de la Guarda (936 km^2) acts to channel the ebb and flow of tides within the gulf. Some of the largest tides anywhere in the world are registered in the northern gulf region, with up to 9 m of vertical displacement (Brusca 1980). The tidal range for the peninsular coast in the southern gulf region is between 2 and 3.5 m. Strong tidal currents characterize the passage between Isla Angel de la Guarda and the peninsular coast (Canal de Ballenas). According to the tales of fishermen, the Canal de Salsipuedes (Leave-If-You-Can Channel) along nearby Isla San Lorenzo is infamous for powerful upwellings that rise up to 3 m above the surrounding water surface and engulf areas up to 0.75 km across (Cannon 1966).

The mean annual surface temperature of gulf waters is 24°C, which is significantly higher than the 18°C norm for coastal waters off the Pacific coast of Baja California (Roberts 1989). Mean average rainfall for the Baja California peninsula is 15.3 cm, although the inner coastal zone along the gulf receives far less. High aridity through much of the year means that the Gulf of California behaves as a huge evaporation basin, estimated to lose nearly 1 m of surface seawater on a yearly basis (Bray 1988a).

Gulf of California's Geological Origins

To clarify what is known about the geological formation of the Gulf of California, and thus shed light on the configuration of the area and the diversity of coastal environments, we provide a summary of the geological history for the entire region. Explanations for the origin of the gulf emphasize the role of plate tectonics, seafloor spreading, and associated fault activity with the evolving gulf. Evidence supporting the proposed history of the gulf derives from geological and paleontological studies in Baja California and mainland Mexico, as well as from geological and geophysical studies in the adjacent deep sea (Gastil et al. 1979; Stock and Hodges 1989; Nicholson et al. 1994; Oskin et al. 2001; Carreño and Helenes 2002; Carreño and Smith 2007).

The Gulf of California geological province can be divided into five regions: (1) the modern subaerial Salton Trough, (2) North Gulf Region, (3) Central Gulf Region, (4) South Gulf Region, and (5) Gulf Mouth Region. Today, the Salton Trough is the only region not covered by marine water. During the region's early marine history, however, it was connected directly with the gulf.

The Baja California peninsula and adjacent Gulf of California evolved in proximity to the Mexican mainland through three phases of development during Miocene to Recent times. The first phase entailed a subduction regime, active from 29 to about 12 Ma along the Pacific coast of Baja California (Hausback 1984). The second phase was associated with a major episode of crustal extension and opening of the proto-gulf (13 to 3.5 Ma) linked to Basin and Range development in western North America (Karig and Jensky 1972; Stock and Hodges 1989). The final and present phase involves the transtensional regime responsible for the present tectonic configuration in the Gulf of California (Zanchi 1994; Mayer and Vincent 1999). This last phase resulted in transfer of Baja California from the North American plate to the Pacific plate.

At about 29 Ma, the East Pacific Rise encountered the Franciscan Deep Sea Trench (Fletcher et al. 2003),

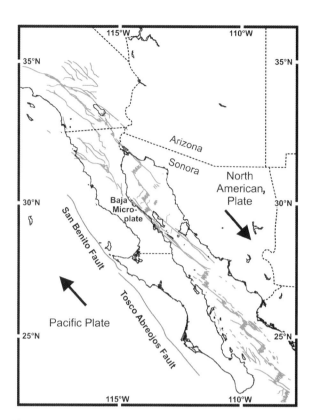

Figure 1.2. Map of Baja California and the Gulf of California showing major tectonic plates, faults, basins, and additional features.

the Pacific plate made contact with North America, and the Farallon–North American subduction margin met the Pacific-Farallon ridge-transform system near present-day Tijuana to form the Pacific–North America transform boundary (fig. 1.2). Starting about 25 Ma, the locus of subduction-related magmatism jumped across the trench from Mexico to a relatively narrow zone near the Pacific coast (Severinghaus and Atwater 1990).

The Mendocino and Rivera triple junctions moved north and south, respectively, lengthening the Pacific–North American transform boundary. Each jump of the triple junction extinguished magmatism progressively farther south along the Late Oligocene–Miocene arc. By the Late Miocene, all of northwestern Mexico was transferred from a subduction margin to a transform margin. The resulting coupling of North America to the Pacific plate caused wrenching at the plate boundary and within the North American plate (Bohannon and Parsons 1995). During the Early to Middle Miocene (24 to 11 Ma), a terrestrial volcanic arc was formed along what is now roughly the axis of the modern Gulf of California (Hausback 1984; Sawlan and Smith 1984). Those volcanic rocks constitute the basement for the Gulf Extensional Province (Gastil et al. 1973). They survive as mountains composed of thick andesite flows bordering the western side of the gulf in peninsular Baja California (fig. 1.3), including the Sierra La Giganta in Baja California Sur near Loreto. Younger volcanism started about 13 Ma in eastern Baja California and within the developing Gulf of California rift (Sawlan 1991).

Late Miocene volcanic centers formed at accommodation zones, and the developing rift became the site of concentrated (Late Miocene) extension leading to the development of the proto-gulf rift (Axen 1995; Oskin and Stock 2003). East–west to east–northeast/west–southwest extension began by 12 Ma, generating normal faulting and tilting of cortical blocks. Extension was accommodated by high-angle normal faults on the surface, but with depth these become low-angle faults known as listric detachment faults (Axen 2000). Surface exposures of this fault type occur along the San Pedro Mártir Fault in northern Baja California and Sierra La Giganta in Baja California Sur, both located along the Main Gulf Escarpment.

Proto-gulf extension took place from about 12 to 6 Ma. By about 5 Ma, extension within the now transtensional plate boundary had rotated by 30° to 50° oblique to the margin. This resulted in the rifting of the Baja California peninsula from North America and development of the modern San Andreas Fault System (Oskin and Stock 2003). Spreading began around the gulf mouth about 5.6 Ma at spreading axes that went extinct as individual spreading centers formed and propagated within the gulf (De Mets 1995).

Basins opened sequentially in the South Gulf Region from southeast to northwest beginning at 5.6 Ma on the María Magdalena Rise, 3.7–3.4 Ma in the Alarcón basin, and 2.1 Ma in the Guaymas basin (Nagy and Stock 2000). Spreading took 3 million years to propagate from the mouth to the Central Gulf Region, and only now may be arriving in the Northern Gulf Region (Nagy and Stock 2000). By the end of the Miocene, the Gulf Extensional Province varied in width from 400 km in the northern gulf to 250 km in the south (Stock and Hodges 1989). The gulf accommodates a maximum of 255 ± 10 km of post-6-Ma dextral slip (Oskin et al. 2001). Pacific–North American transform motion was switched from the California borderland to the gulf region and the San

Andreas Fault System between 6.3 and 4.7 Ma (Oskin et al. 2001).

Marine Incursions

Exploratory wells in the Northern Gulf Region reveal a 12-Ma date for the oldest marine incursion within the Proto-Gulf of California (Helenes et al. 2005). According to Winker and Kidwell (1996), marine inundation reached the San Gorgonio Pass about 6.5 Ma (but never crossed through to the Los Angeles basin and Pacific Ocean). During the Early Pliocene, the Colorado River deposited sediment into the Salton Trough, and over time the rift developed as a lacustrine basin (Winker and Kidwell 1996). In the northern Salton Trough, fossil and radiometric data from strata in the Fish Creek Gypsum suggest that a marine incursion occurred at 9.5–8 Ma (Dean 1996; McDougall et al. 1999). Biostratigraphic evidence restricts the age for marine incursions recorded in the overlying Imperial Formation to 6.5–6.3 Ma. For the southern Salton Trough, microfossils are also consistent with a Late Miocene or earliest Pliocene (< 6 Ma) onset of marine incursion.

Biostratigraphy of a diatomite deposit near San Felipe shows that the oldest marine deposits at this location were deposited between 6.0 and 5.5 Ma (Boehm 1984) and that deposition occurred in a relatively deep basin (200 m). The associated microfossils indicate high productivity in a subtropical to temperate environment (Helenes et al. 2005). South of San Felipe, the Puertecitos Formation consists of Miocene and Pliocene strata with sedimentological characteristics of shallow-marine intertidal to subtidal environments and a fossil assemblage interpreted as having formed under warm and shallow conditions similar to today (Stock et al. 1996). Farther south on Isla Tiburón, paleontological and radiometric data from marine strata provide Late Miocene to Early Pliocene dates, indicating that the marine incursion occurred at 6.5 Ma (Oskin and Stock 2003) or somewhat earlier (Carreño and Smith 2007).

Localized basins that show the episodic influence of marine water were formed in Bahía de Los Angeles, Bahía Las Animas, Isla Angel de la Guarda, and the San Lorenzo Archipelago (Carreño and Smith 2007, fig. 35c). The stratigraphic sequence in the San Lorenzo Archipelago consists of evaporitic and clastic deposits representing transgressive and regressive marine episodes. At Sierra Las Animas, the age

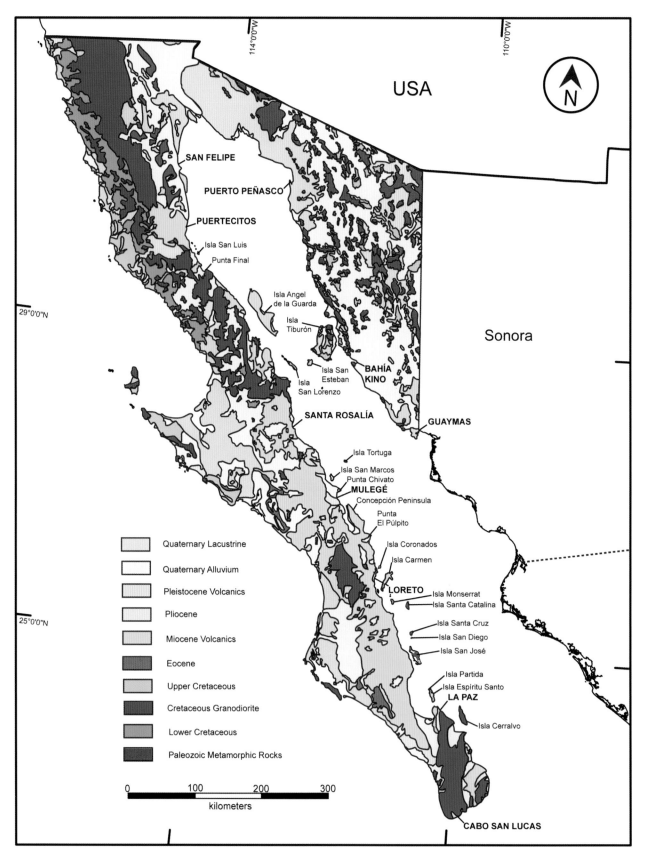

Figure 1.3. Geological map of the Baja California peninsula and adjacent coastal zone of Sonora modified from the *Carta Geológica de México* by Salinas-Prieto et al. (2007).

of andesitic lava underlying the sedimentary rocks is 7.8 ± 0.2 Ma, therefore the age of the marine sedimentary sequence is dated from approximately 7.8 to 5 Ma (Escalona-Alcázar et al. 2001).

The Santa Rosalía basin is a composite of several sub-basins separated by tilted basement blocks. Coarsening-upward cycles of marine to nonmarine deposits dominate. The basal Boleo Formation has not yielded fossils. However, Holt et al. (2000) obtained a $^{40}Ar/^{39}Ar$ date from a tuff in the Boleo Formation, which they combined with magnetostratigraphy to estimate an age of 7.09–6.93 Ma for the base of the formation and 6.27–6.14 Ma for the top of the unit. Alluvial fans, fan deltas, and littoral to neritic marine environments are represented in Late Miocene and Early Pliocene sediments of this basin.

Within the basal Boleo Formation, gypsum forms a continuous deposit that is flat to gently dipping. Dome or mound structures up to 200 m in diameter are common and exhibit outwardly concentric dips reaching 30° (Bales et al. 2001). Evaporitic conditions are regarded as the origin of the gypsum in this area. The overlying Tirabuzón, Infierno, and Santa Rosalía Formations, in contrast to the underlying Boleo Formation, are composed of richly fossiliferous marine sediments. The base of the Tirabuzón Formation, which sits unconformably on Boleo beds, is dated as 5.3 Ma, representing an Early Pliocene age (Carreño 1981).

Late Miocene marine strata near the gulf mouth show that the Pacific Ocean had access to the South Gulf Region (fig. 1.4) between 9 and 8 Ma, possibly due to local extensional stress and crustal thinning (Stock and Hodges 1989). There are no marine deposits in the Gulf Mouth Region that correlate with more northerly marine deposits in a simple way, but marine strata suggest that a proto-gulf opened in the Middle and Late Miocene during extension and that the present (modern) Gulf of California developed during the Pliocene following rifting of Baja California from the North American plate. As outlined above, we now know that regional subsidence related to crustal extension in the Miocene and transtensional activity since the Middle Pliocene produced fault-delimited basins that filled with sediments during those times.

Helenes and Carreño (1999) proposed a seaway through the central part of the Baja California peninsula, providing a connection between the open Pacific Ocean and the proto-gulf during the Miocene. The idea is supported by mixed cool-tropical characteristics of fossil assemblages and the distribution of

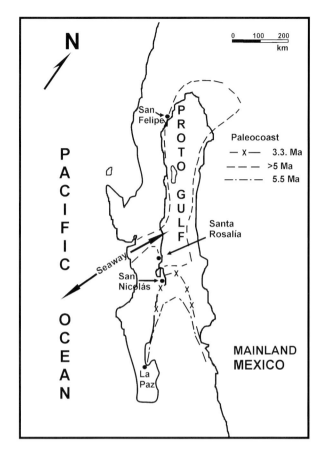

Figure 1.4. Proposed position of a seaway across the Baja California peninsula during the Late Miocene to Early Pliocene (modified from Ledesma-Vázquez 2002).

marine deposits in the north and nonmarine deposits in the central and southern parts of the Gulf of California (Helenes and Carreño 1999; Ledesma-Vázquez 2002; Carreño and Smith 2007). A more northern seaway is unlikely on the basis of the tropical and subtropical characteristics of fossil assemblages (Smith 1991b; Winker and Kidwell 1996; Helenes and Carreño 1999). Work by Bohannon and Parsons (1995) indicated that a seaway is precluded from the north or from the west by the presence of a plutonic belt.

The precise location of the seaway is unclear, but considering the extent of the Peninsular Batholith, it was probably south of the Sierra San Pedro Mártir, perhaps around El Barril on the gulf coast, and conceivably connected to the Pacific Ocean via Bahía San Ignacio on the Pacific side (fig. 1.4). At San Ignacio, the Middle to Late Miocene San Ignacio Formation represents an ideal candidate for part of the seaway. The fossil assemblage from this formation occurs in marl and includes endemic Tertiary Caribbean and South American species (Smith 1984; Carreño and

Smith 2007, fig. 35b). Above the sedimentary units, the Esperanza basalt erupted at about 10 Ma under low-energy subaqueous conditions (Sawlan and Smith 1984).

Due to vertical motions near the tip of the Baja California peninsula and mainland Mexico, Late Miocene subsidence (10–5 Ma) resulted in marine deposits of a relatively deep-water nature, now exposed north of San José del Cabo (Carreño 1992). Subsequently in the Pliocene, the area was uplifted and remained above sea level. Phylogeographic studies on the terrestrial fauna inhabiting the Cabo Trough suggest that the tip of the peninsula was separated for a long time from mainland Mexico. One possible scenario is that the Los Cabos Block migrated northeast along the La Paz Fault approximately 50 km to its present location relative to the rest of Baja California, which implies that the block remained largely isolated. Another scenario suggests that the Los Cabos Block did not separate from the mainland as an isolated landmass, but remained connected to the rest of the peninsula.

According to Fletcher et al. (2000), evidence points toward the second scenario, but with a particular geological history that provides a clearer understanding of the evolutionary history of the species occurring in the region. During the Early to Middle Miocene (16–10 Ma), shallow seaways were present on the western margin of the Los Cabos Block that promoted marine deposition. After the region was uplifted, the eastern margin received marine deposits in the Middle Miocene (Carreño 1992) that reached bathyal depths by the Late Miocene to Pliocene.

Oceanography

The Gulf of California has been described as a region of high primary productivity (Alvarez-Borrego and Lara-Lara 1991), but it is also widely recognized as an area critical for reproduction and an important nursery for many nektonic species such as shrimp, sardine, and shark. With a relatively small shelf area of approximately 20,000 km² above the 100-m isobath, the gulf's coastal zone also serves as a rich habitat for diverse benthic marine invertebrates specialized for intertidal to shallow subtidal life. Without efficient mixing of gulf waters and the fertilization of phytoplankton and zooplankton at the base of the food chain, this great diversity of life would be much impoverished. The extremely high productivity characteristic of the Gulf of California is a result of strong turbulent kinetic energy that supplies the euphotic zone with a constant source of nutrients (Gilbert and Allen 1943; Santamaría-del-Angel et al. 1994a, 1994b; Lavín et al. 1995).

As a rule, gulf winds display strong seasonal patterns. Over offshore regions, strong northwesterly winds prevail from November to May, bringing cool-dry air under winter conditions. Under summer conditions, the winds blow from a southerly direction, bringing hot, often moist air (Merrifield and Winant 1989; Alvarez-Borrego and Lara-Lara 1991). These extreme seasonal variations in atmospheric conditions have been documented with discrete meteorological and hydrographic data (Lavín et al. 1995) and have been confirmed using satellite imagery (Badan-Dangon et al. 1985; Santamaría-del-Angel et al. 1994a, 1994b).

Based on wind patterns, Roden (1964) predicted that upwelling events within the Gulf of California occur on the eastern coast during winter conditions (under northwesterly winds) and on the western coast during summer conditions (under southeasterly winds). This system has been checked only against oceanographic data for the existence of winter upwelling events. Ocean-color imagery shows that for summer in the central and the southern parts of the gulf, pigment values are generally higher on the eastern than on the western sides of the gulf (Santamaría-del-Angel et al. 1999). During the winter season, plumes of high chlorophyll concentration may be seen extending from the eastern side of the gulf toward the Baja California coast. In general under winter conditions, the gulf reflects high-pigment values (Gendrop-Funes et al. 1978; Alvarez-Borrego 1983; Valdez-Holguín 1986; Alvarez-Borrego and Gaxiola-Castro 1988; Bray 1988b; Merrifield and Winant 1989; Paden et al. 1991; Cajal-Medrano et al. 1992; Lara-Lara et al. 1993; Millán-Núñez et al. 1993; Santamaría-del-Angel et al. 1994a) as a direct result of the high turbulent kinetic energy generated during that season of the year.

Chlorophyll *a* levels indexed from ocean-color imagery under winter conditions register high concentrations with clear pattern structure (fig. 1.5A), and this is also true under summer conditions (fig. 1.5B). In general, these levels are the inverse of sea-surface temperature (SST) patterns. SST imagery (fig. 1.5C) shows values between 17 and 23°C, with the maxima found in the south and minima in the northern mainland area. Possibly due to eddy circulation that stimulates upwelling, small to medium structures of low

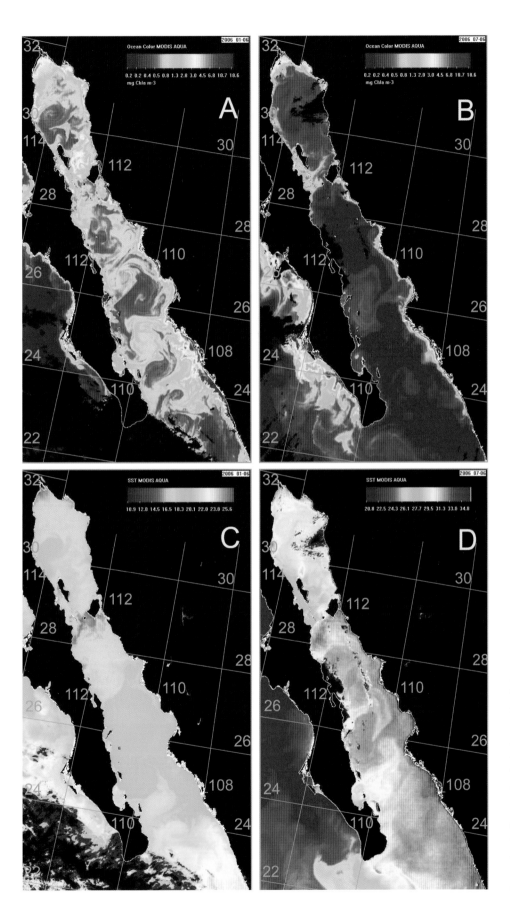

Figure 1.5. Winter and summer conditions of visible spectral radiometry (chlorophyll *a*) and infrared radiometry (sea-surface temperature) imagery by Aqua/Modis for the Gulf of California: (A) chlorophyll *a*, January 6, 2006; (B) chlorophyll *a*, June 6, 2006; (C) sea-surface temperature, January 6, 2006; (D) sea-surface temperature, June 6, 2006.

SST can be seen, generally from the eastern coast to the western coast of the gulf. Similarly, development of isotherm structures form as complex structural patterns with one or more low-SST locations, generally along the coast of mainland Mexico.

During summer conditions, the Gulf of California sustains low-pigment surfaces (Santamaría-del-Angel et al. 1994a, 1994b) associated with a well-developed thermocline (Simpson et al. 1994). Isotherm maps for summer conditions typically show SST values between 28 and 31°C. In general, the "lower" SST values (28–29°C; fig. 1.5D) appear off the coast of Baja California under two patterns. The first pattern forms as random "low" SST pools, whereas the second pattern is much like a thermal front with isotherms arranged quasi-parallel to the coast. These structural patterns suggest one-cell circulation that brings upwelled waters from west to east. Isoline maps for surface pigments during some summer days show "high" pigment values (such as 0.20 and 0.21 mg/m^3) with one-cell patterns that are clear and inverse to the isotherm maps. The majority of high-pigment values occur off the western coast with isolines that are parallel to the coast.

To summarize, the results of the ocean-color and SST analyses show "low" SST (between 28 and 30°C) in summer conditions on the western coast, while during winter conditions the lowest SSTs (17 to 19°C) are found in the region south of the larger islands. Taking an isotherm of 30°C for summer and 19°C for winter, we see a displacement of the cold plume from west to east nearer to the south under summer conditions and from east to west closer to the north for winter conditions. Regions of high pigment value are located off the western coast during the summer (0.20 and 0.21 mg/m^3) and off the eastern coast during the winter (2.0 to 4.9 mg/m^3). Displacement patterns are denoted west to east during the summer and the inverse during the winter. The pattern of the upwelling effect on the surface-space distribution of the phytoplankton biomass during winter conditions is clear (fig. 1.5), as is the pattern of high-pigment values under summer conditions.

Under winter conditions with winds from the northwest (Merrifield and Winant 1989), the Surface Eastern Tropical Pacific (SETP) water mass and the Subsurface Subtropical (SS) waters, both of which are oligotrophic-hot waters, occur only at the gulf mouth. Under summer conditions with winds from the southeast (Merrifield and Winant 1989), SETP and

SS waters enter into the Gulf of California (Alvarez-Borrego and Schwartzlose 1979). Castro et al. (1994) asserted that invasion of oligotrophic-hot water during the summer flows into the gulf close to the eastern coast in the form of structural plumes like circulation cells with very low concentrations of pigment. This condition is visible in the ocean-color imagery for summer conditions (Santamaría-del-Angel et al. 1994a).

Alvarez-Borrego and Schwartzlose (1979) reported that during the summer of 1957 (an El Niño year) they encountered strong incursions of SETP and SS water masses. Using data for temperature and salinity, Torres-Orozco (1993) calculated the volume of SETP gulf waters for the period between 1939 and 1986. He reported a clear seasonal and interannual pattern. A minimum volume was estimated for the start of the winter season, but the volume grew in increments through to the summer, with the maximum reached during the first half of winter conditions. Maxima volumes have occurred during El Niño years, with the highest value achieved in 1983.

Badan-Dangon et al. (1985) observed that under winter conditions low-SST plumes were very well developed, originating at a point along the continental coast and crossing over the gulf to reach the peninsular coast. These plumes were moved by the wind away from the continental coast, crossing the gulf to the opposite side, where they divided into feathers that spread out and returned toward the continental coast. There are no reports on plume behavior under summer conditions with regard to superficial temperatures.

As shown by surface pigment values, displacement of plume flukes based on SST data are due to superficial currents in the Gulf of California. Several studies reported that a summer net flow toward the north hit the gulf's east coast (Alvarez-Borrego 1983; Marinone and Ripa 1988; Torres-Orozco 1993) and produced superficial cyclonic circulation cells (Bray 1988a). Under winter conditions, the net flow shifts toward the south (Alvarez-Borrego 1983; Marinone and Ripa 1988) with anticyclonic circulation cells (Bray 1988a). These anticyclonic cells have been detected using Advanced Very High Resolution Radiometer imagery (Badan-Dangon et al. 1985).

The upwelling system in the Gulf of California is associated with the presence of SETP water masses brought into the gulf. Under summer conditions, SETP water sits in the gulf like a hot "cap" that masks the effect of upwelling and decreases nutrient transport to the euphotic zone. This occurs in a similar manner as the El Niño mask known to disrupt upwelling in regions such as Peru. The winds that cause the upwelling front off the coast of Peru may be intensified during an El Niño event, but the accumulation of hot, nutrient-poor water forms a cap with a deep nutricline. The result is that upwelling of nutrient-poor water blocks the growth of phytoplankton. In summary, for summer conditions characterized by water with a oligotrophic-hot cap the intrusion of SETP water into the Gulf of California insures that upwelling does not bring high concentrations of nutrients. The upwelling does exhibit a greater nutrient concentration than occurs in surface water, as reflected by light increments in pigment concentration. On a global scale, however, this increase is small.

Under winter conditions, SETP water is flushed out of the Gulf of California, and the system behaves like a classical upwelling system in that cold water brings with it a high concentration of nutrients readily apparent in a high pigment concentration. Alvarez-Borrego (1983) and Alvarez-Borrego and Lara-Lara (1991) documented that the 2.5 μM isoline of phosphates occurs in shallow water (near the 100-m isobath off the coast in the South Gulf Region) to create conditions of very high fertilization in the euphotic zone.

Sedimentation Patterns

The principal suppliers of clastic sediments to the Gulf of California today are rivers draining the Sierra Madre Occidental of southern Sonora, Sinaloa, and Nayarit, although prior to 1935 the Colorado River delivered large amounts of sediment in the north (Brusca and Bryner 2004; Brusca et al. 2005). Far less sediment comes from Baja California, where the watershed is close to the gulf coast, although abundant alluvial fans, fan-deltas, and extensive deep-sea fans are developed around the tip of the peninsula. In the Central Gulf Region, huge populations of plankton in the highly productive surface waters provide a heavy rain of pelagic debris, which makes up about half of the sediment in several of the deep marine basins (fig. 1.6). Ponds of relatively pure diatomaceous ooze fill some depression in these basins. Hemipelagic muds have high organic carbon content, especially where the seafloor intersects an oxygen-minimum zone that occurs between 300 and 800 m in the central and southern gulf. In those areas where reducing conditions prevail, diatomaceous oozes occur on the

continental slope as thin laminae alternating with muddier laminae (Calvert 1966).

Biogenic carbonate in slope and basin sediments is composed of benthic and planktonic foraminifera and coccolithophorids. Pteropod, molluscan, echinoderm, and algal (rhodolith) debris are common on the continental shelf and banks, but little is transported into deeper water. Planktonic foraminifera, especially *Gobigerina bulloides* and *Neogloboquadrina dutretrei*, are abundant in upwelled waters (Pride 1997) but are not always abundant in the sediments due to carbonate

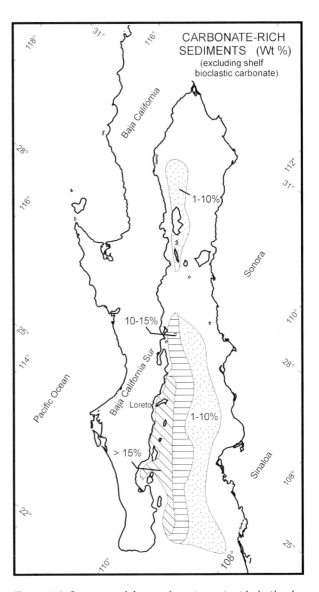

Figure 1.6. Core-top calcium carbonate content in bathyal sediments from the Gulf of California (excluding shelf bioclastic carbonate). The main components of the sediments are benthic and planktonic foraminifera and coccolithophorids (modified from Douglas et al. 2007).

dissolution in the water column and on the seafloor. In the Guaymas basin, a comparison of sediment-trap results from the sediment directly beneath suggests that about 70–80 percent of the carbonate flux was destroyed (Pride 1997). The carbonate content of surface samples declines with increasing water depth, especially in depths associated with intermediate waters and the oxygen-minimum zone. Coccoliths are more prevalent in oligotrophic waters of the western side (Macias-Carranza 1999; Ziveri and Thunell 2000) and contribute an important part of the carbonate found in the sediments. A band of high carbonate values (> 10 percent) occurs along the peninsula margin, with the highest concentrations located south of Isla Carmen (fig. 1.6). Foraminiferal muds and sands are common and are mostly accumulating in areas with low sedimentation rates. The areas of high carbonate lie beneath low productivity areas (pigment concentrations < 1.5 mg/m^3) and along the pathway of water as it exits the gulf. These more saline and nutrient-poor waters generated by evaporation in the northern gulf during the winter (Bray 1988b; Bray and Robles 1991) flow just below the surface mixed layer (Castro et al. 2000). Water from the Central Gulf Region may be an important factor contributing to low productivity on the western side of the gulf (Douglas et al. 2007).

From south to north, the biogenic production of carbonates ranges from oligotrophic-mesotrophic coral-dominated shallow-water areas through mesotrophic-eutrophic red algal–dominated inner-shelf areas in the Central Gulf Region to molluscan-bryozoan eutrophic inner- to outer-shelf environments (Halfar et al. 2004). Therefore, corals, calcareous red algae (rhodoliths), and mollusks are the primary limestone producers (Halfar et al. 2001, fig. 13). Modern and ancient counterparts for these components are covered in chapters 4 and 5 (for corals and coral reefs), chapters 6 and 7 (rhodoliths), and chapters 8 and 9 (bivalved mollusks). As explored in this volume, the various biological components of the present and past carbonate factory tend to express distinct geographical patterns of distribution within the Gulf of California. To highlight recent developments concerning only one dynamic aspect of the carbonate factory, Hetzinger et al. (2006) demonstrated how acoustic mapping of the modern shallow shelf south of Isla San José led to the discovery of sizable rhodolith banks. In their study of the Pliocene-Pleistocene Arroyo Blanco basin on Isla Carmen, Eros et al. (2006) found that massive amounts of crushed rhodolith debris are buried in a vast carbonate ramp that occupies a former embayment.

Submarine hydrothermal activity is known to occur in diverse tectonic settings throughout the world's oceans, including the Gulf of California. High-temperature hydrothermal venting, where heat is transferred from the lithosphere to the ocean, is associated with active plate boundaries such as spreading centers and fracture zones. Resulting from regional extension as well as crustal thinning, hydrothermal vents were activated within the Gulf Extensional Province as early as the Miocene. Active vents are found from the intertidal and shallow subtidal zones (Forrest et al. 2005) to deep-sea ridge segments such as the Guaymas basin. There, copious discharge of geothermal liquids and gases occurs through fractures in rock and soft sediments along linear trends generally parallel to fault zones. Deep sites like the Guaymas basin are very similar to the axial rift valley of midoceanic ridges. Chemo-autotrophic hydrothermal-based ecosystems, highly specialized with endemic micro- and macrofauna, are located at these sites. Hydrothermal venting in shallow areas affects the local marine ecology, as the diversity and abundance of biota on rocky habitats adjacent to diffuse venting sites appear to be enhanced, particularly in fish and epifaunal filter-feeding invertebrate assemblages, whereas infaunal animals are significantly less diverse and abundant in areas of active venting through soft sediment (see chap. 12).

2 Peninsular and Island Rocky Shores in the Gulf of California

David H. Backus, Markes E. Johnson, and
Jorge Ledesma-Vázquez

Excluding the 180-km-wide gap at the opening to the Gulf of California, Mexico meets the Pacific Ocean along shores that stretch 4,500 km in length. Shallow lagoons protected by sandy barrier islands are especially common along low-lying Pacific shores in Oaxaca, Sinaloa, and Baja California Sur. Taking into account the deep incursion made by the Gulf of California, the ocean frontiers of western Mexico expand by 66 percent to enfold a coastline about 7,500 km in length. Sea cliffs in this realm are highly concentrated on the gulf coast of the Baja California peninsula and its associated islands (Gutiérrez-Estrada and Ortiz-Perez 1985). Transfer of peninsular territory from the North American plate to the Pacific plate occurred through rift and strike-slip tectonics that propagated much of the rugged topography on the gulf's western shores. Shore cliffs include patches of granite (or granodiorite) and metamorphic rocks, extensive andesite and other volcanic rocks, as well as sporadic marine limestone (Beal 1948; Fenby and Gastil 1989). Most of the igneous rocks and some limestone layers are highly resistant to erosion.

From an altitude of 9,000 m, travelers who fly commercial routes to and from Loreto, La Paz, or Los Cabos in Baja California Sur are presented with a spectacular coastal landscape. During the months from November through June, the sky is often cloudless. Mountain ranges descend in rank from between 700 and 900 m to intersect narrow coastal plains interrupted by dry washes from interior basins. Against this dramatic backdrop, the gulf's surface waters often are perturbed by great swells that look like small ripples from high above. The swells advance southward in long fronts pushed by strong winter winds that blow for days at a time. Wind-driven waves energize water that surges directly or obliquely onto all the beaches or rocky shores standing in the way. Hence, the physical geography of the place is wild and dynamic on multiple levels.

The distribution of rocky shores is a topic that can be treated in great detail based on data collected from satellite images (Wadge and Quarmby 1988). Reconnaissance from low-flying aircraft is a thrilling experience, but satellite technology provides many advantages for the otherwise desk-bound explorer. The field of view covered by satellite sensors is vast and permits instantaneous comparison of adjacent regions at high resolution. The layout and quantification of rocky shores as classified by different rock types, analysis of coastal discordance (extent of embayments), and interpretation of sea-cliff profiles are areas of interest to physical geographers (Woodroffe 2002). Variations in sea-cliff profiles relate to rock strength and local history of sea-level changes, factors best studied in place on the ground. Even so, clues to these aspects of rocky shores may be gleaned from satellite images.

Viewed from any distance above the Baja California peninsula, the immense sparseness of the desert landscape is evident but masks the rich diversity of marine life known to occur in adjacent gulf waters (Brusca 1980; Brusca et al. 2005). Tolerance to wave shock is the most important factor controlling the distribution of intertidal to shallow, subtidal marine invertebrates and algae (Knox 2001). Thoroughly mixed and nutrient-rich waters are efficiently delivered to wave-exposed rocky shores, whereas sheltered shores tend to undergo some degree of water stratification, siltation, and reduced nutrient supply. Thus, the variety and abundance of marine life encountered at low tide around the gulf islands and peninsular coast depend on wind and water-circulation patterns linked to nutrient supply through upwelling that have persisted over thousands of years. To a lesser degree, the species of marine organisms associated with rocky shores are related to the kind of rocks that form sea cliffs and their relative hardness. Fieldwork by marine ecologists and biogeographers informs these topics, which are summarized as the second major component of this chapter.

Evaluating Satellite Images

Satellite images generated by the Enhanced Thematic Mapper (ETM+) and Advanced Spaceborne Thermal Emission and Reflection Radiometer (ASTER) sensors were the primary resources consulted on shorelines in the Gulf of California. Images available from Google Earth (www.earth.google.com) also were checked for the amount of topographic expression on rocky shores and the width of fringing sandy beaches. ETM+ images cover about 180 km^2 at a resolution of 30 m/pixel, whereas ASTER images cover only 60 km^2 but at a higher resolution of 15 m/pixel. A series of seven ETM+ images were downloaded from the Global Land Cover Facility at the University of Maryland. In addition, 26 ASTER images were acquired that cover overlapping areas with few gaps for the gulf coast of the Baja California peninsula from the Colorado River delta to Cabo San Lucas. The remote sensing software ENVI ver. 4.0 by ITT Visual Information Solutions was used to modify available ETM+ and ASTER satellite data for all coastal segments of interest. For each image, the best three-band combination for visualizing rocks as different from one another as granite and limestone was used to process a false-color image. In the case of ETM+ data, bands 7,4,1 were combined to create the composite image to showcase coastal limestone deposits. In addition, resolution of the resulting three-band combination was improved to 14.5 m/pixel using a Gram-Schmidt Spectral Sharpening function that employed the ETM+ panchromatic band. Bands 8,3,1 proved to be the best three-band combination when using ASTER data (repository on CD, this volume).

The resulting images were compared for accuracy to existing geological maps and published descriptions. The northern state of Baja California enjoys geological coverage at a scale of 1:250,000 (Gastil et al. 1973, 1975). With an area of 1,223 km^2, Isla Tiburón is the largest island in the Gulf of California; it was described and mapped at a scale of 1:150,000 (Gastil et al. 1974; Gastil and Krummenacher 1977). More detailed geological mapping at a scale of 1:70,000 is an exception, as performed by McFall (1968) around Bahía Concepción in Baja California Sur. Geological

maps showing considerably less detail for the regions surrounding the gulf were compiled at scales of 1:1,000,000 by Beal (1948) and 1:1,528,900 by Fenby and Gastil (1989). Short descriptions of the geology for the gulf islands were provided by Anderson (1950), Gastil et al. (1983), and Carreño and Helenes (2002).

Original sources on the marine biology of rocky shores in the Gulf of California were culled from the published literature, with special reference to foundation studies such as those by Steinbeck and Ricketts (1941) and Brusca (1980).

Physical Geography of Rocky Shores

A false-color satellite image that exemplifies the inherent cragginess of rocky shores in the Gulf of California is an ETM+ image centered about Loreto, Baja California Sur (fig. 2.1). The image covers the gulf coast from the landing at San Bruno 24 km north of Loreto to Playa La Ballena 48 km south of town. The larger islands shown in full are Islas Coronados, Carmen, Danzante, and Monserrat. Particularly striking is the way limestone deposits show up as whitish beige patches on most of the islands.

Coastal Classification and Quantification

Analysis of peninsular shores via satellite images commenced at El Golfo de Santa Clara at the end of Highway 40 on the Sonoran side of the border below the Colorado River delta and concluded at Land's End on Cabo San Lucas. To facilitate tabulation, the gulf coast was divided into segments with an average coastal length of 140 km. Five categories of rocky shores were defined by lithotype: granodiorite, andesite, volcanic rocks, metamorphic rocks, and limestone. This classification is strongly correlated with geologic age (see geological map in fig. 1.3). The metamorphic rocks mostly are Paleozoic in origin, the granodiorite is Cretaceous, the andesite is mostly Miocene, the other volcanic rocks are Pleistocene, and the limestone is Pliocene and/or Pleistocene in age (Anderson 1950; Gastil et al. 1975; Fenby and Gastil 1989). In addition, two kinds of soft shores were considered: beaches composed of cobble- to sand-size materials and mud flats composed of silt- to clay-size materials. In total, 1828.8 km on the gulf shores of the Baja California peninsula were categorized by rock type or sedimentary grade and measured to the nearest tenth of a kilometer with respect to coastal length (table 2.1).

The political boundary at 28°N does not divide the

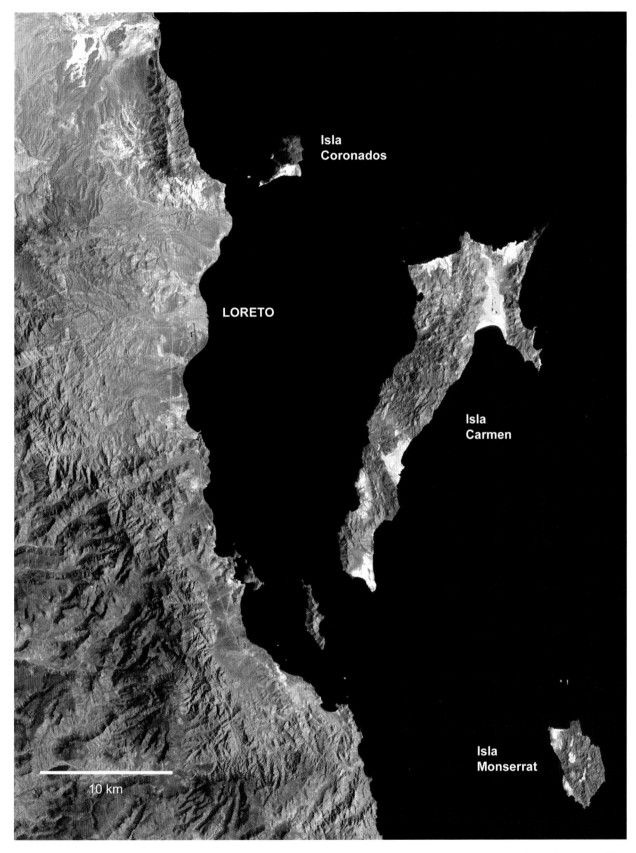

Figure 2.1. False-color satellite image for the coastal region around Loreto and the Bahía Loreto National Park in Baja California Sur. The field of view for this image is approximately 72 x 45 km. In this band combination (7,4,1), limestone deposits on some of the islands clearly show up as beige patches. Note the pale blue coloration for the salt lagoon on Isla Carmen. ETM+ image acquired from the Pan-American Center for Earth and Environmental Studies. Original data set was collected by NASA LANDSAT Program, LANDSAT TM scene LT4035042008916110, L1G, USGS, Sioux Falls, June 10, 1989.

Table 2.1. Peninsular gulf coast of Baja California and the gulf islands (G = granite; A = andesite; V = young volcanics; M = metamorphic rocks; L = limestone; C = cobble and/or sandy beach; and S = silty and/or muddy shore). All data tabulated in kilometers (km).

Region	G	A	V	M	L	C	S	Total
Baja California	59.5	81.1	3.4	32.4	44.6	359.0	110.1	690.1
Baja California Sur	61.5	204.6	39.7		121.2	705.5	6.2	1138.7
Sub-total	121.0	285.7	43.1	32.4	165.8	1064.5	116.3	1828.8
Northern Islands	44.4	196.7	47.7	31.4	12.4	200.5	33.0	566.1
Southern Islands	128.6	232.7	27.0	2.5	43.2	103.7	7.0	544.7
Sub-total	173.0	429.4	74.7	33.9	55.6	304.2	40.0	1110.8
Totals	**294.0**	**715.1**	**117.8**	**66.3**	**221.4**	**1368.7**	**156.3**	**2939.6**
Coastal percentages	9.1%	24.3%	4%	2.3%	7.5%	46.6%	5.3%	100%

peninsula equally in terms of gulf frontage (dotted line in fig. 2.2), but comparisons between the northern state of Baja California and the southern state of Baja California Sur are useful. The northern state accounts for only 38 percent of the peninsular gulf shoreline, while the southern state commands 62 percent. The southern state enjoys greater access to the gulf due to major coastal incursions at Bahía Concepción and Bahía La Paz, as well as the broad tip of the peninsula and its termination with the granite sea arches at Cabo San Lucas. The amount of granodiorite exposed as rocky shores on the peninsular gulf coast is roughly equal in the two states, but the amount of limestone in the south is more than 2.5 times that found in the north. Andesite is the dominant rock exposed as sea cliffs on the gulf coast in both states (see fig. 1.3).

Eighteen islands that reside above 28°N in the northern gulf were studied through satellite images, and 15 islands were studied in the southern gulf. In total, 1,110.8 km on the perimeter of gulf islands were categorized by rock type or sedimentary grade and measured to the nearest tenth of a kilometer with respect to coastal length. The two largest islands, Tiburón and Angel de la Guarda, are situated in the northern gulf (fig. 2.2), but many medium-size islands are found in the southern gulf. Hence, the composite perimeters for the northern and southern gulf islands are closely matched at 566.1 and 544.7 km, respectively (tables 2.2, 2.3). Granitic islands, such as Santa Catalina, are more common in the southern gulf. The amount of limestone on island shores in the southern gulf is 3.5 times that found on island shores in the northern gulf. Andesite is the dominant coastal rock type found in both sets of islands.

Taking the peninsular gulf coast and all the larger islands into account, the full expanse of shoreline approaches 3,000 km. The breakdown in distribution shows that soft sediments (cobble- to silt-size materials) form 52 percent of the coast and rocky shores make up the remaining 48 percent. Nearly 25 percent of all shores are formed by andesite, while 9 percent are granite and 7.5 percent are limestone in composition (table 2.1). Fully half the muddy and sandy shores in the northern state of Baja California are concentrated between Puertecitos and the influential delta of the Colorado River (fig. 2.2). One-quarter of the cobble-to-sandy shores in Baja California Sur are concentrated in two areas, on the eastern side of Bahía Concepción and within Bahía La Paz. Alluvial fans that coalesce to form bajadas are extensively developed on the flank of the west-tilted fault blocks of the Concepción Peninsula marginal to Bahía Concepción. Upper Oligocene to Lower Miocene sandstone strata form much of the escarpment north of La Paz as far as Punta Coyote (fig. 2.2), but these rocks are overlain by extensive Pleistocene cover near the shore or are easily eroded at sea level to supply beaches and the enormous sand spit protecting La Paz harbor.

The principal factors responsible for the distribution of rocky shores in the region are tectonic in nature. Granitic coasts are more common on gulf islands than adjacent peninsular shores, because the granodiorite was brought to the surface in coherent fault blocks from the deep basement. Coastal granite on the gulf is equivalent to granite found inland 1,700 m below the surface near 28°N at Las Tres Vírgenes and dated to 84 Ma (López-Hernandez et al. 1994). The famous sea arches at Cabo San Lucas

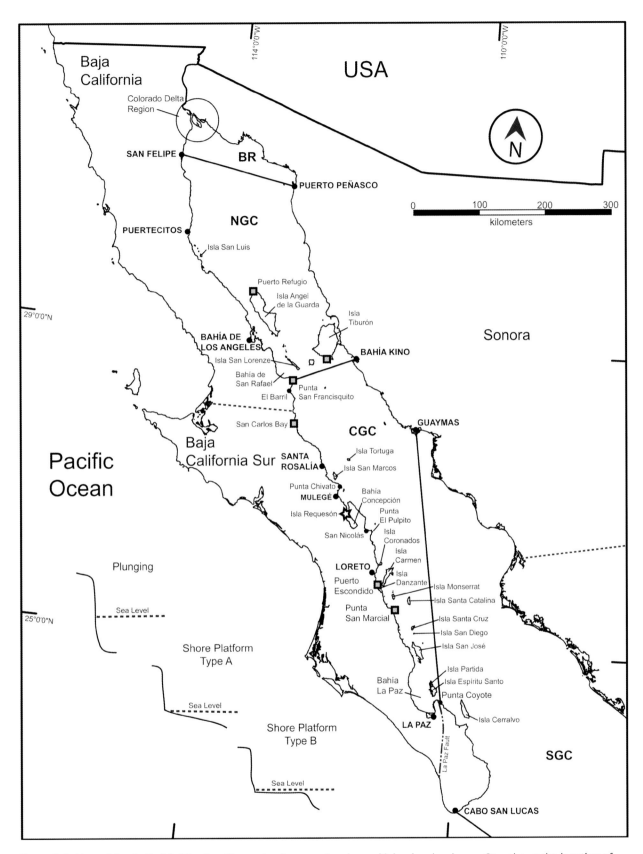

Figure 2.2. Map of the Gulf of California with emphasis on peninsular and island rocky shores. Star shows the location of census study by Hayes et al. (1993). Squares show location of rocky-shore collection sites visited by Steinbeck and Ricketts (1941). Inset illustrates variations in cliff profiles common in the Gulf of California.

Table 2.2. Island shores, northern Gulf of California (G = granite; A = andesite; V = young volcanics; M = metamorphic rocks; L = limestone; C = cobble/sandy beach; and S = silty/muddy shore). All data in kilometers (km).

Island name	G	A	V	M	L	C	S	Total
Isla Miramar		5.3				1.3		6.6
Isla Salvatierra		1.8						1.8
Isla Encantada		3.5						3.5
Isla San Luís			8.3			1.6	2.9	12.8
Isla San Luís Gonzaga			6.4			3.4		9.8
Isla Mejia	0.3	0.2		2.5		5.3		8.3
Isla Granito	2.4					0.5	0.3	3.2
I. Angel de la Guarda	10.3	109.6		4.8	12.4	69.1	7.9	214.1
Isla Estanque = Pond		0.5				4.7		5.2
Isla Coranodo = Smith	1.5			6.2		16.4		24.1
Isla La Ventana	2.7			1.8		1.6		6.1
Isla Tiburón	8.3	12.2	29.5	8.6		89.2	19.8	167.6
Isla Partida		5.5				0.7		6.2
Isla Raza			3.5					3.5
Archipelago S. Lorenzo	18.9	36.1		1.9		5.7	2.1	64.7
Isla San Esteban		22.0		5.6		1.0		28.6
Totals	**44.4**	**196.7**	**47.7**	**31.4**	**12.4**	**200.5**	**33.0**	**566.1**

Table 2.3. Island shores, southern Gulf of California (G = granite; A = andesite; V = young volcanics; M = metamorphic rocks; L = limestone; C = cobble/sandy beach; and S = silty/muddy shore). All data in kilometers (km).

Island name	G	A	V	M	L	C	S	Total
Isla Tortuga			11.1			2.2		13.3
Isla San Marcos		22.7			5.6			28.3
Isla Ildefonso			6.9					6.9
Isla Coronados		4.1	9.0			4.8		17.9
Isla Carmen		76.6			25.8	4.8		107.2
Isla Danzante		16.3				1.5		17.8
Isla Monserrat		16.7			3.3	2.9		22.9
Isla Santa Catalina	30.3			2.5		4.5		37.3
Isla Santa Cruz	17.1					0.3		17.4
Isla San Diego	4.0							4.0
Isla San José	16.6	13.6			8.5	46.6		85.3
Isla Francisco		6.8				4.2		11.0
Isla La Partida		33.7				2.2	2.2	38.1
Isla Espíritu Santo	7.6	42.2				17.9	4.8	72.5
Isla Cerralvo	26.4			26.6		11.8		64.8
Totals	**102.0**	**232.7**	**27.0**	**29.1**	**43.2**	**103.7**	**7.0**	**544.7**

date to roughly the same Cretaceous age (Sedlock et al. 1993), and the same granodiorite is widely represented inland on the huge Los Cabos Fault block (see fig. 1.3). Andesite is an igneous rock, also enriched in silica like granite, but extrusive in origin. From 25 to 5 Ma through much of the Miocene, andesite was deposited in surface flows punctuated by volcanic breccias and tuffs that grew collectively to 4,000 m in thickness over western Mexico. Generally attributed to a catchall stratigraphic unit called the Comondú Group (McFall 1968), these layers were faulted during the rifting of the Baja California peninsula from the Mexican mainland beginning about 14–10 Ma (Stock and Hodges 1989; Lyle and Ness 1991). Concentration of andesite shores reaches 50 percent on the peninsular coast between Punta Prieta at Mulegé and Punta El Pulpito 207 km to the south. Isla Angel de la Guarda is the second largest island in the Gulf of California (fig. 2.2). With more than 50 percent andesite shores, it offers more coastal kilometers of this rock type than any other island. Islands such as Isla Danzante in the southern gulf show higher concentrations of coastal andesite, but are much smaller. Young islands such as Isla San Luís in the north and Isla Tortuga in the south retain volcanic cone structures and exhibit extensive basalt or other volcanic shores (tables 2.2, 2.3).

Global to local changes in sea level are a secondary factor that influenced the placement of rocky shores in the Gulf of California, particularly with regard to limestone. More so in southern gulf areas than to the north, low-lying coastal lands were subject to flooding during highstands in sea level that took place during the Pliocene between 5.5 and 1.8 Ma. Subsequent sea-level retreat exposed limestone beds deposited in shallow basins along the peninsular coast near Loreto, San Nicolás, Punta Chivato, Santa Rosalía, El Barril, Punta San Francisquito, and Bahía de San Rafael (fig. 2.2). One of the longest and most continuous sections of limestone shore on the peninsular gulf coast extends for 8 km on the Ensenada El Mangle north of Loreto (Johnson et al. 2003). Some of the larger islands such as San José, Monserrat, Carmen, San Marcos, and Angel de la Guarda also accumulated shallow limestone deposits during the Pliocene. With more than 25 km of limestone coast in discontinuous segments, Isla Carmen has the most extensive shore exposures of this lithotype in the Gulf of California.

Smaller marine incursions with associated carbonate deposits occurred during later Pleistocene times around 400,000, 325,000, 200,000, and 125,000 years ago (Ortlieb 1991), but were restricted to narrow terraces that underwent subsequent uplift. Pleistocene terraces eroded from Pliocene limestone are especially well exposed on the eastern side of Isla Carmen at Arroyo Blanco (Eros et al. 2006), where the 12-m terrace forms a plunging sea cliff that extends laterally for 1.25 km.

Evolution of Coastal Discordance

Shores aligned as a rigid front against the sea define a concordant coast typically controlled by geological structures. In contrast, features striking perpendicular to the regional trend of a shoreline make a discordant coast (Woodroffe 2002). Recessed beaches flanked by rocky headlands represent an example of a discordant coast on a local scale. Headlands may exhibit erosional decay in the form of sea caves, sea arches, and sea stacks. Given sufficient time, a concordant coast will evolve into a discordant coast under conditions of normal shore erosion abetted by changes in sea level that drown drainage systems already situated in low-lying areas. In this regard, coastal discordance is a measure of geomorphic maturity. Like the Red Sea between Africa and Arabia, however, the Gulf of California represents a geologically young seaway. Evidence of coastal discordance is present in the Gulf of California but bears distinct tectonic imprints. Bahía Concepción in Baja California Sur varies from 5 to 10 km in width, but penetrates 40 km inland (see fig. 2.2). The deep embayment is an example of a pull-apart basin expressed as a half-graben. It is attributed to the east–west extension that more widely initiated rifting of the Baja California peninsula from the Mexican mainland starting in the Late Miocene at about 10 Ma (Ledesma-Vázquez and Johnson 2001). The southern part of the Baja California peninsula is cut from side to side by the La Paz Fault, separating the Los Cabos block or Pericú Terrane on the tip of the peninsula (Sedlock et al. 1993). Movement of the block on the fault was mostly vertical and took place from the Late Cretaceous to Miocene, although sufficient oblique movement on associated faults gave Bahía La Paz its discordant shape prior to establishment of transform faults in the Gulf of California at about 3.5 Ma.

Many midsize islands in the Gulf of California are elongate or oval in shape with concordant rocky shores parallel to the long axis. Islands entirely or partially composed of granite easily fall into this category, including Cerralvo, San José, San Diego, Santa Cruz, and Santa Catalina. Andesite-dominated islands, such as Monserrat, Danzante, San Marcos, and San Lorenzo, also are elongate but distinctively more knotted in outline and less concordant. The difference is due largely to the even erosion of plutonic rocks that are more uniform in composition, as opposed to extrusive rocks with variations in composition and hardness from one level to another (i.e., andesite interbedded with tuffs and breccias). The most discordant shorelines in the Gulf of California occur on Isla Espíritu Santo and its close neighbor Isla La Partida. The eastern side of Espíritu Santo is composed of granodiorite, for example, but tilted andesite layers of variable thickness and hardness strike off to the western side of the island. This configuration promoted the erosion of several long inlets oriented perpendicular to the western coast of the island. The rake-like headlands that separate the many inlets on Espíritu Santo make the island one of the most photographed in the gulf.

Islas Carmen and Angel de la Guarda are intermediate in coastal discordance. Both exhibit irregular coastlines typical of andesite shores, but also feature sizable bays anchored by headlands set at oblique angles to the principal island axis. These islands are sufficiently large to host multiple fault patterns that control placement of embayments. Bahía Salinas on the northeastern side of Carmen is influenced by a major graben within which the island's extensive salt lagoon is situated (Anderson 1950). Ensenada del Pulpito on the eastern coast of Angel de la Guarda is bordered by faults that parallel the outer sides of the embayment.

Sea-Cliff Profiles

Woodroffe (2002) depicted three kinds of rocky-shore profiles (see inset, fig. 2.2). They include plunging cliffs that fall vertically into the water, cliffs with a Type A ramp that gradually slope into the water, and cliffs with a Type B platform that assume a subhorizontal rim near sea level. Variations involve the presence or absence of talus at the foot of sea cliffs and the extent to which talus forms a seaward-extending apron. Depending on wave energy and tidal range, plunging cliffs also may acquire a wave-cut notch. Figures 2.3–2.6 illustrate the sea-cliff lithotypes and profiles most characteristic of the gulf region. More than 70 percent of Isla Monserrat's coastline is formed by Miocene andesite eroded to form high cliffs footed by a Type B platform clear of talus (fig. 2.3). At low tide, it is possible to walk around the island, keeping to the rock platform much of the way. Pleistocene

volcanic rocks form high cliffs and sea stacks on the northern coast of Isla Coronados (fig. 2.4), which receives strong wave activity during the winter months. A clean Type B platform is exposed around the sea stacks at low tide, but deep plunging cliffs also exist on the northern coast. Soft limestone cliffs with an attached ramp covered by talus dominate the southern shores of Isla Coronados (fig. 2.1). A laterally persistent talus pile constructed mostly by andesite cobbles keeps seawater away from these low Pleistocene cliffs. Hence, limestone rocky shores are poorly registered for Isla Coronados (table 2.3).

Granodiorite sea cliffs of various heights account for 40 percent of the coastline of Isla Cerralvo, although often broken by steep delta fans carrying outwash from intersecting canyons. Evenly spaced, vertical joints typical of granite that result in the weathering of rounded features on the rock face are evident on Cerralvo from well offshore (fig. 2.5). The same granite cliffs distinguish Los Cabos on the nearby peninsular coast (see fig. 1.3), although broad beaches rich in silica sand insulate much of the cliff line from seawater. The narrow neck of land that reaches the arches at Land's End features plunging granite sea cliffs familiar to the many tourists who visit the area aboard glass-bottomed boats.

Stratified Pliocene limestone tends to erode as plunging sea cliffs, as along the Ensenada El Mangle north of Loreto (fig. 2.6). Thrill-seeking geology students who have jumped from the 15-m high cliffs on this part of the peninsular coast swear to the absence of Type A ramps or Type B platforms at the base of the cliffs and the scarcity of submarine tallus. The red cliffs on the periphery (center right, fig. 2.6) clearly demonstrate, however, that the original carbonate sediments transgressed a ramped andesite shore during the Pliocene.

Rocky-Shore Biology

A rich profusion of marine life populates cliffs and tide pools on the granodiorite shores of Bahía San Lucas near the exposed tip of the Baja California peninsula (Zwinger 1983). Harsh conditions in the high intertidal zone are tolerated by the periwinkle gastropod (*Littorina*). Rock surfaces in the upper part of the mid-intertidal are crowded by balanid barnacles (*Balanus*), while neritid gastropods (*Nerita scabricosta*) typically confine themselves to shaded vertical cracks in the rock. The lower part of the mid-intertidal is

Figure 2.3. (top left) View looking northwest on the western shore of Isla Monserrat demonstrating an andesite sea cliff with a Type A platform exposed during low tide. Cliff height is approximately 8 m. Photo credit: Markes E. Johnson. **Figure 2.4.** (bottom left) View looking northwest on the eastern shore of Isla Coronados demonstrating volcanic sea cliff with sea stacks on a Type A platform exposed during low tide. Center sea stack is approximately 8 m in height. Photo credit: Markes E. Johnson. **Figure 2.5.** (top right) View looking west toward the eastern side of Isla Cerralvo demonstrating typical granite sea cliffs. Cliff height is approximately 15 m. Photo credit: Markes E. Johnson. **Figure 2.6.** (bottom right) View looking west toward the peninsular coast on the Ensenada El Mangle (25 km north of Loreto) demonstrating plunging limestone sea cliffs (white) that adjoin andesite sea cliffs (red, to right). Photo credit: Markes E. Johnson

frequented by chitons (*Chiton virgulatus*), thaid gastropods (*Purpura pansa*), and sun stars (*Heliaster kubiniji*). Calcarous tubes belonging to a serpulid worm (*Serpulorbis margaritaceus*) may be thickly encrusted on rock ledges at this level. The lower intertidal zone is occupied by medium-sized colonies of stony corals such as *Porites panamensis* and *Pocillopora elegans* together with colonial anemones (*Palythoa*), all firmly cemented on granite. Pink in color and bush-like in growth, coralline red algae (*Jania*) thrive here, too.

Farther north, the andesite cliffs on Isla Monserrat are entirely barren of marine encrusters (fig. 2.3), as seawater rarely splashes these rocks. Sally lightfoot crabs (*Graspus graspus*) are the undisputed masters of the wide platform that protrudes from the base of the cliffs. The roughness of the platform gives the crabs protective cover. There is little evidence of other life

before reaching the outer margin of the rock platform, where balanid barnacles reside. It is much the same around the sea stacks on Isla Coronados (fig. 2.4). Small colonies of the coral *P. panamensis* occur sporadically in the lower range of the intertidal zone along these coasts, but are more abundant in the adjacent shallows.

Plunging limestone shores, as on the Ensenada El Mangle (fig. 2.6), attract other specialists. Similar Pliocene limestone forms a low headland at Punta Cacarizo on the Punta Chivato Promontory north of Mulegé (fig. 2.2). The purple sea urchin (*Echinometra vanbrunti*) is found in large numbers on these rocks, each snuggly fitted within a cuplike depression worn in the comparatively soft limestone (Johnson 2002a). Competing for space with the sea urchins is an olive-green colonial anemone (*Palythoa ignota*) that occupies areas up to 0.5 m in diameter. *Porites panamensis* thrives in scattered colonies in the shallows along the entire front of Punta Cacarizo and reaches into the lower intertidal zone on the backside of the point on the edge of protected lagoons.

Variations on Windward and Leeward Rocky Shores

In one of the most detailed census studies performed on rocky-shore biotas in the Gulf of California, Hayes et al. (1993) collected data on diversity and abundance patterns in the intertidal zone around Isla Requesón in Bahía Concepción. This small, oval-shaped island has a circumference of slightly more than 2 km, 95 percent of which is formed by andesite sea cliffs. The cliffs are footed by a Type A ramp covered with boulder- to cobble-sized talus fully submerged by high tides with a range of 2.75 m. Marine invertebrate and algal populations were canvassed using a 25-×-25-cm quadrat with a sample interval of 2.5 m through transects with an average length of 17.5 m across the intertidal zone to the shallow subtidal zone. A total of 66 stations were sampled. Transects spaced in parallel lines but perpendicular to the coast separated by 20 m tested the potential for small-scale variation. Transect pairs separated by about 200 m recorded large-scale variations on opposite sides of the island.

The survey resulted in the registration of 30 species of macroorganisms with marked differences for toleration to wave exposure. Rocky shores and talus rubble with a north-facing exposure are extensively encrusted by coralline red algae below a clear zone of balanid-dominated barnacles. The bivalve *Arca pacifica* is abundant in the mid-intertidal zone, typically anchored by coralline algae. Scattered colonies of *P. panamensis* and large mats of colonial anemones (*Palythoa*) occupy the lower intertidal zone. In contrast, the barnacle *Tetraclita affinis* is more common on rocky shores and associated talus with a south-facing exposure. The bivalve *Ostrea palmula* occurs abundantly in the mid-intertidal zone, often accompanied by the gastropod *Crucibulum spinosum*. Green algae (*Ulva*) and the sponge *Verongia aurea* occur sporadically in the lower intertidal zone. By observation of physical conditions, Hayes et al. (1993) concluded that the northern coast biota thrived in a windward, high-energy environment, while the southern coast biota enjoyed greater protection in a leeward, low-energy environment. Lateral variation within the two settings proved to be minimal. No other studies of comparable scope are known with which to compare the

Isla Requesón data, but the results of locality-specific surveys are informative regarding wave exposure and faunal composition.

The Ricketts-Steinbeck Survey of 1940

The research trip to the Gulf of California by Edward Ricketts and John Steinbeck in 1940 is regarded as the first modern scientific foray into the gulf. Partly philosophical, but mainly biological in scope, the original edition of the popular book that chronicled the trip (*Sea of Cortez: A Leisurely Journal of Travel and Research*) by Steinbeck and Ricketts (1941) included 40 plates that illustrate common invertebrates found along gulf shores, as well as an extensive appendix not found in the modern edition.

The appendix summarizes the structure of the near-shore invertebrate faunal within the Gulf of California at the time and includes an annotated list of all the invertebrates encountered on the trip with relevant references. More recently, a verbatim transcript of Ricketts' original notes from the gulf trip was included in a compilation of papers edited by Rodger (2006). Ricketts' original notes provide more in-depth information not available in the original travelogue about each locality where invertebrates were collected. Using these resources, we compiled a species list (table 2.4) for six rocky shore localities collected by Steinbeck and Ricketts (1941). Two localities, Punta Refugio at the northern end of Isla Angel de la Guarda and Bahía San Carlos near the Baja peninsular coast, are unprotected localities fully exposed to gulf winds and wave action. The remaining four collections were from protected sites at the southern end of Isla Tiburón near the Sonoran coast of the gulf, and the coastal embayments at San Francisquito, Puerto Escondido, and Punta San Marcial on the Baja peninsula (fig. 2.2).

The data in table 2.4 should be viewed with some care. In contrast to Hayes et al. (1993), where the surface of the rocky shore was the focus, Ricketts and company extended their collecting to the bottoms of and the interstices between the rocks on the shore. The attention of the collectors also was not focused in the same way at each site. For instance, Ricketts made no mention of marine algae, whatsoever. Also, the collection of specimens was not made under identical conditions at each locality (Steinbeck and Ricketts 1941). Each tide was of a different duration and magnitude, while the time spent collecting specimens was not necessarily the same at each location. Each sample

Table 2.4. Marine invertebrates collected at exposed (2) and protected (4) rock-shore localities in the Gulf of California by Steinbeck and Ricketts (1941). Species names are those originally used in the 1941 edition.

List of Species	Puerto Refugio	San Carlos Bay	Isla Tiburón	San Francisquito	Puerto Escondido	Punta San Marcial
Porifera						
Aaptos van namei	X				X	
Geodia mesotriaena	X					X
Hymeniacidon spp.	X					
Leucetta losangelensis	X			X		
L. heathi	X					
Spirestrella sp.			X			
Tethya aurantia			X		X	
Coelenterata						
Astrangia pedersenii			X		X	X
Porites porosa	X		X			X
Bryozoa						
Bugula neretina						X
Flustra sp.	X					
Lagenipora erecta						X
Lichenspora sp.						X
Membranipora sp.					X	
Porella sp.			X			
Scrupocellaria diegensis				X		
S. scruposa						X
Echinodermata (Asteroidea)						
Heliaster kubiniji	X	X	X	X	X	X
Acanthaster ellisii					X	
Leiaster teres					X	
Pharia pyramidata					X	X
Phataria unifascialis	X	X	X	X	X	X
Linkia columbiae	X	X	X	X	X	X
Mithrodia bradleyi	X				X	
Astrometis sertulifera	X	X	X	X	X	X
Othilia tenuispina	X	X	X	X	X	X
Oreaster occidentalis					X	
(Ophioroidea)						
Ophiocoma aethiops			X			
O. alexandri	X					X
O. spiculate	X		X		X	
Ophiactis simplex			X			
Ophionereis annulata	X					
Ophioderma teres	X			X	X	

Table 2.4. *continued*

List of Species	Puerto Refugio	San Carlos Bay	Isla Tiburón	San Francisquito	Puerto Escondido	Punta San Marcial
(Echinooidea)						
Eucidaris thouarsii	X	X	X	X	X	X
Echinometra vanbrunti						X
Arbacia incisa	X					
Centrostephanus cornatus	X		X			X
Centrechinus mexicanus						X
(Holothuroidea)						
Holothuria arenicola	X		X		X	
H. impatiens	X				X	
H. lubrica	X				X	
H. rigida					X	
Stichopus fuscus					X	
Neothyone gibbosa					X	
Euapta godeffroyi					X	
Arthropoda (Crustacea)						
Aruga sp.	X					
Elasmopus sp.		X				
Pontharpinia sp.		X				
(Isopoda)						
Cirolana hartfordi			X			
Ligyda exotica		X			X	
Paranthura sp.			X			
Rocinella aries				X		
(Schizopoda)						
Nyctiphanes simplex						X
Archeomysis sp.						X
Mysidopsis sp.						X
(Cirripedia)						
Tetraclita stalactifera		X		X	X	
Chthamalus anisopoma					X	X
Tetraclita squamosa			X			
(Decapoda)						
Palaemon ritteri	X		X			
Crangon (Alpheus) sp.						X
Panulirus inflatus						X
P. interruptus				X		
Calcinus californiensis		X				
Clibanarius digueti			X	X	X	X
Petrolithes gracilis		X				
P. hirtipes	X	X				X
P. nigrunguiculatus		X		X	X	

Table 2.4. *continued*

List of Species	Puerto Refugio	San Carlos Bay	Isla Tiburón	San Francisquito	Puerto Escondido	Punta San Marcial
Pisonella tuberculipes			X			
P. sinuimanus					X	
(Brachyura)						
Dromdia larraburei	X					
Stenorynchus debilis			X			
Podochela latimanus	X					
Eucietops lucasii	X					
E. panamensis		X				
Pitho sexdentata						X
Anaptychus cornutus						X
Microphyrus platysoma	X					
Platypodia rotundata	X					
Glyptoxanthus meandricus		X				
Leptodius occidentalis	X	X				
L. cooksoni		X			X	
Xanthodius hebes						X
Panopeus bermudiensis					X	
Eurypanopeus planissimus	X		X			
Pilumnus gonzalensis	X	X	X			
Eriphia squamata		X				
Graspus graspus	X	X	X	X	X	X
Geograspus lividus				X		
Pachygraspus crassipes				X		
Mollusca (Pelecepoda)						
Barbatia reeveana					X	
Fossularca solida	X	X	X		X	
Fugleria illota	X	X				
Navicula mutabilis						X
Isognomon anomioides	X		X		X	
I. chemnitziana	X	X				
Pinctada fimbriata					X	
P. mazatlanica					X	
Spondylus sp.(limbatus?)					X	
Anomia peruviana					X	
Brachidontes multiformis	X	X				
Lithophaga aristata	X					
Volsella capax					X	
Carditamera californica	X	X	X	X	X	
Chama squamuligera						X

Table 2.4. *continued*

List of Species	Puerto Refugio	San Carlos Bay	Isla Tiburón	San Francisquito	Puerto Escondido	Punta San Marcial
Gastropoda						
Conus princeps						X
Fasciolaria princeps					X	
Engina ferruginosa	X				X	
Columbella fuscata	X					
Nitidella guttata		X			X	X
Acanthina lugubris	X	X		X		
Coralliophila costata					X	
Phyllonotus nigritus				X		X
Thais biserialis			X			
T. centriquadrata			X			
T. tuberculata					X	
Cypraea annettae					X	
Strombus galeatus					X	
Cerithium sculptum	X					X
Aletes squamigerus						X
Crepidula onyx					X	
C. squama		X				
Crucibulum imbricatum					X	
C. spinosum					X	
Nerita berhardi		X			X	
N. scabricostata			X			
Neritina picta					X	
Acmaea atrata	X		X			
A. dalliana	X			X		
A. mesoleuca	X					
A. pediculus			X			
Callopoma fluctuosum	X		X			
Tegula impressa	X					X
T. mariana			X		X	
T. rugosa		X				
Tegula sp.						X
Diadora alta			X			
D. inequalis	X		X		X	
Fissurella rugosa	X		X			
Tethys californica(?)	X					
Dolabella californica(?)					X	
Berthella plumula	X					
Aegires sp.	X					
Tridachiella diomedea			X			

Table 2.4. *continued*

List of Species	Puerto Refugio	San Carlos Bay	Isla Tiburón	San Francisquito	Puerto Escondido	Punta San Marcial
Onchidium lesliei	X	X				
Haminoea strongi	X	X				
Polyplacophora						
Acanthochitona exquisita	X	X			X	X
Chiton virgulatus	X	X	X		X	X
Ischnochiton clathratus	X				X	
I. tridentatus	X	X	X	X	X	X
I. limaciformis			X			
I. conspicuus		X				
Nuttallina cf. *allantophora*	X					
Cephalopoda						
Octopus bimaculatus	X	X				
Chordata (Tunicates)						
Amaroucium californicum	X		X			
Didemnum carnulentum	X			X		
Eudistoma sp.	X		X		X	X
Pyura sp.					X	
Ascidia sp.	X				X	
Cystodytes dellechiajei					X	
Botrylloides diegensis				X		
Pyura sp.					X	
Ascidia sp.	X				X	
Species Totals	**69**	**38**	**45**	**24**	**67**	**43**

reflects only one day of one season of one particular year. It also is likely that collecting was not done strictly from rocky surfaces but from any protected pockets of sediment found within the rocky shore. Therefore, these data are considered to be more informative with respect to higher-level groupings of species (phylum or class level) and less informative when considering the absence or presence of individual species at a given locality.

Of the six localities collected, the sites with the greatest diversity were Puerto Refugio (exposed) with 69 species and Puerto Escondido (protected) with 67 species. The lowest number of species (24) was collected at Punta San Francisquito. As found in table 2.4, the invertebrate groups with the broadest distribution among the localities were the asteroids (sea stars) and the polyplacophoran mollusks (chitons). In fact, the top predator of the rocky shore, the sea star *Heliaster kubiniji*, is among the three most common

species (all echinoderms) found in the gulf (Steinbeck and Ricketts 1941; Brusca 1980). The other species noted for their abundance by Steinbeck and Ricketts (1941) were the pencil urchin *Eucidaris thouarsii* and the sea cucumber *Holothuria lubrica*, also found on rocky shores.

Of the 162 species collected from all six of the rocky shore localities, 35 species occurred exclusively at the exposed localities of Puerto Refugio and Bahía San Carlos, while 77 species were found exclusively at the protected localities. A total of 49 species were found at both protected and unprotected sites. On average, a straight comparison of the totals for exposed and protected localities suggests that there is about 1.5 times as many species at a protected rocky shore as found at an exposed rocky shore. If we compare only the species found exclusively at either exposed or protected sites, however, this ratio changes from approximately 1:1.5 to about 1:2. With these ratios in mind, we looked at the various invertebrate groups represented by the data, compared their distribution to the general pattern found in the aggregate data, and looked for notable deviations.

Within the Phylum Bryozoa, organisms that thinly encrust a variety of rock and biological surfaces in the gulf are represented by only eight species. Only one species is recorded from an exposed locality, however, while seven species were found exclusively at the protected sites.

The Phylum Echinodermata is represented by 28 species, over half of which occur at both exposed and protected localities. As noted above, asteroids (sea stars) have the broadest distribution of any group. They also include several species that were exclusively found at the protected locality of Puerto Escondido. A similar pattern appears for the soft-bodied holothurians (sea cucumbers). Combined, these two groups have eight species that appear to be exclusive to protected rocky shores. In contrast to the echinoderms, the arthropods include many species that are exclusive to either exposed or protected environments. Only nine of the 44 species collected were found in both environments. Distribution of the Brachyura (true crabs) is of particular note. Of the 20 species collected, eight species were found exclusively at exposed localities as compared to seven species found exclusively at protected sites. No other invertebrate group in our study shows a distribution so relatively weighted toward the exposed localities.

Of all the invertebrate groups collected by Steinbeck and Ricketts (1941) the mollusks were the most diverse, with 64 species from the six rocky shore localities reviewed, the largest number (41 species) coming from Class Gastropoda (snails). The distribution of mollusk species among the exposed and protected locations follows the general distribution pattern found for the invertebrates as a whole, with two interesting exceptions. Polyplacophorans (chitons) are the most widely distributed mollusk among the six localities studied. All but one species from this low-profile, armor-plated group were collected at exposed localities. The other molluscan subgroup with a notable distribution is the shell-less gastropods (opisthobranchs), of which four of six species were found exclusively at the more exposed localities. One of the two exceptions is the beautiful sea hare (*Tridachiella diomedea*), also known as the Mexican dancer.

Finally, there are the ascidians (tunicates), an unusual group that is technically a chordate but due to its primitive form and ecological characteristics it is usually included among the invertebrates. Of the nine species found in the gulf by Steinbeck and Ricketts, four species were exclusive to the protected localities, while the other five species were found in both exposed and protected environments.

The Steinbeck and Ricketts (1941) data are intriguing and suggest that future studies on modern leeward and windward faunas in the Gulf of California will provide greater insight into the distribution of invertebrates on rocky shorelines.

Regional Patterns in Rocky-Shore Biodiversity

The importance of rocky shores to the diversity of both animals and marine algae within the Gulf of California and the greater Pacific Coast of North America is a well-established idea (Ricketts et al. 1953; Brusca et al. 2005). In general, the distribution of benthic invertebrates on the coasts of the Pacific region is controlled by the intensity of wave surge, the amplitude of the tide in the region (exposure time), and the substrate or type of bottom. The Gulf of California is an elongate cul-de-sac that has an extremely strong winter wind (El Norte) and a tidal range that increases northward toward the delta of the Colorado River. When the relationship between species numbers and substrate is considered, general patterns can be discerned. Diversity is highest on reefs, stable shores, and shallow marine shelves composed of roughly surfaced, easily eroded rock types. Unstable beaches of sand or rolling cobbles and shorelines carved out of hard rock types with smooth surfaces have the lowest diversity. Areas with a mix of substrates types are more diverse than coastal regions with a single substrate type (Brusca et al. 2005).

A recent compilation of the invertebrates within the Gulf of California by Brusca et al. (2005) lists 4854 species. They estimated this number represents only half the actual standing diversity of invertebrates in the gulf. Of known species, over 88 percent (4299 of 4854 species) are listed by Brusca et al. (2005) as benthic (bottom-dwelling) organisms. It is likely that an overwhelming majority of these benthic species are associated with rocky shores. To date, the available data have not been compiled to support this assertion. However, an accounting of how these species are distributed among rocky shores, tidal flats, or sandy beaches should be possible when the Gulf of California Biodiversity Database becomes available (Brusca, personal communication 2006). For now, our best view of the distribution of all invertebrates in the gulf is through a standardized four-zone system originally derived from the distribution of fishes in the gulf and used in the database (fig. 2.2). Using the gulf coast of the Baja peninsula as a reference, the Southern Gulf of California Region (SGC) extends from Cabo San Lucas to Punta Coyote (north of La Paz). Punta San Francisquito marks the northern end of the Central Gulf of California Region (CGC), and the boundary between the northern end of the Northern Gulf of California Region (NGC) and the Colorado River Delta Bioreserve Area (BR) is found at San Felipe.

According to Brusca et al. (2005, table 9.4), the central gulf region (3293 species) is marginally more diverse than the southern gulf region (3,113 species), followed by the northern gulf region (2,258 species) and the Colorado River Delta Biosphere region (1,050 species). As in our previous analysis of the Steinbeck and Ricketts (1941) rocky shore collections, most of the invertebrate groups follow the general trend of the aggregate data with a few interesting exceptions. The Porifera (sponges) are modestly more diverse in the northern gulf region. In contrast, the numbers of crabs (Anomuran and Brachyuran) and echinoderms are noticeably higher for the southern gulf region. In the future, a more sophisticated analysis of the data likely will provide better information about the distribution of benthic invertebrates in the Gulf of California.

Another significant group of organisms that colonize rocky shores within the Gulf of California are the marine algae. Faced with the same environmental conditions as the sessile benthic invertebrates, marine algae are dependent on what the ocean brings them. Research within the gulf and elsewhere suggests that seaweed distribution is also controlled by water temperature, oxygenation of the water, and the supply of nutrients (Villalard-Bohnsack and Harlin 1992; Aguilar-Rosas et al. 2000; Cruz-Ayala et al. 2001).

Regional studies have identified 140 species of marine algae in the area from Puertecitos to the north, 108 species of Rhodophytes (red algae) within Bahía de Los Angeles, and 126 species of seaweed identified in the Bahía La Paz area (Aguilar-Rosas et al. 2000; Cruz-Ayala et al. 2001; Pacheco-Ruiz and Zertuche-Gonzaléz 2002). Historical records indicate that, in all, 151 species have been identified within Bahía de

Los Angeles, while 284 species of seaweed have been found at one time or another in Bahía La Paz. These numbers may suggest there is an increase in diversity from south to north within the gulf, but a closer look at the relationship between species diversity, substrate type, and the seasonal cycle in seaweed diversity suggest a different story.

The data from the northern gulf come from three rocky-shore localities near Puertecitos and three sandy-beach sites from further to the north. An average of 79 species was collected from the rocky shores, with a high of 85 seaweed species examined at Puertecitos. In contrast, the sandy-beach localities yielded an average of only 26 species (Aguilar-Rosas et al. 2000). In Bahía La Paz, the most diverse locality yielding 66 species of seaweed was a rocky shore at Calerita near the mouth of the bay, southeast of Isla Espíritu Santo (Cruz-Ayala et al. 2001). The Calerita locality included a mix of tidal pools, rocky platform, and sandy bottom that lends support to the idea that heterogeneous substrates support larger numbers of species than homogenous ones. Cruz-Ayala et al. (2001) also pointed out that the seaweeds collected exclusively at Calerita are indicative of stable or mature communities. Repeated seasonal collections at all localities in Bahía La Paz, Bahía de Los Angeles, and the northern gulf region near Puertecitos show a fluctuation in species diversity throughout the year that is important to consider. Each of the three studies cited above found an inverse relationship between water temperature and seaweed diversity within the gulf (Aguilar-Rosas et al. 2000; Cruz-Ayala et al. 2001; Pacheco-Ruiz and Zertuche-Gonzaléz 2002). There also seems to be a large variation in species that recruit available substrates within all regions of the gulf on a yearly basis. This is true particularly in the northern region, where the annual range in temperature (15–29°C) is greatest (Cruz-Ayala et al. 2001).

Clearly, the rocky shore is an important Gulf of California environment for many organisms. However, the long-term patterns of rocky-shore recruitment and use by invertebrate animals and marine algae are still not fully understood. We echo the call of Brusca et al. (2005) for continued research on rocky shores in the Gulf of California due to their importance to regional biodiversity.

3 Pliocene and Pleistocene Development of Peninsular and Island Rocky Shores in the Gulf of California

Markes E. Johnson and Jorge Ledesma-Vázquez

Former rocky shorelines in the Gulf of California have a record that dates back 5.5 Ma or more, illustrating conditions when the sea level was higher and the gulf reached farther north than today. The gulf coast of peninsular Baja California and associated islands is a vast natural laboratory for the study of rocky shores abandoned above sea level. Prevailing desert conditions prohibit the growth of lush vegetative ground cover, except for rare seasons when rainfall exceeds the norm, as most recently witnessed in 2004–2005. Arid conditions make it easy for geologists to identify sharp contacts between different sorts of rocks that mark a former rocky shore and to trace those contacts uninterrupted across the landscape. Commonly expressed by geologists as "walking the outcrop," tracking such contacts on foot is an activity readily converted to a line trace on a topographic map.

The result of such an exercise is the production of an accurate paleogeographic map that reconstitutes the original dimensions and contours of a long-abandoned coastline. In summarizing descriptions from earlier studies and collecting new examples of former shorelines from locations throughout the Gulf of California, we sketch the rich geological and paleontological history that preceded the arrangement of rocky shores and rocky-shore life found in the gulf today (chap. 2). Because the configuration of the proto-gulf during the Miocene remains obscure (see chap. 1), the story told here covers the interval from earliest Pliocene to later Pleistocene times (5.5 Ma to about 125 Ka). During the Middle Pliocene at about 3.5 Ma, the gulf underwent a dramatic change in tectonic regime that still affects the ongoing development of the region and continues to alter the coastline through relatively high rates of coastal uplift. Gulf of California tectonics played out against a backdrop of global sea-level changes, and the trade-off between these factors in the retirement of old rocky shores is an important consideration. The distribution of marine organisms living on former rocky shores also provides another important insight on the continuity of life in the Gulf of California.

Definitions and Physical Parameters

The essential geological contacts that denote former rocky shores are unconformities in which sedimentary rocks with marine fossils rest against older configurations often composed of igneous rocks or metamorphic rocks. By definition, unconformities are abrupt and they signify an interval of missing time between the final accumulation of one rock type and the initial accumulation of another that touches directly on it. Unconformities around the Gulf of California are most typical where Pliocene or Pleistocene limestone laps against Miocene andesite (dated from 19 to 13 Ma) or older Cretaceous granodiorite (dated from 99 to 78 Ma). Granodiorite is an intrusive igneous rock that forms domed bodies congealed from the contents of deeply buried magma chambers. Cretaceous magma chambers in Baja California originated many kilometers below the surface of the Earth. Before Pliocene or Pleistocene sedimentary rocks could be deposited around the granodiorite, the domes were exhumed by deep erosion. Andesite is an extrusive igneous rock that forms layers resulting from surface flows of volcanic material. During the extensional phase of the gulf during the Miocene, andesite layers were stretched, fractured, and tilted. As a result, Pliocene sedimentary rocks later deposited around andesite fault blocks define an angular unconformity. The contact between sedimentary rocks and intrusive igneous rocks like granodiorite constitutes a nonconformity.

Pliocene rocky shores in the gulf region fit one of three geographic parameters. Entire paleoislands are preserved intact with circular contact boundaries that represent a coastline formerly open to the sea on all sides. Unless tectonically disturbed later, limestone beds surrounding paleoislands retain their original seaward dip of about 6°. The inverse situation finds flat-lying sedimentary rocks preserved in small basins related to the drowning of former inland valleys or depressions almost entirely surrounded by high ground. The sedimentary rocks make an abutment unconformity at the basin margins, where they terminate against the flanks of older surrounding hills. Marine fossils from such basins show that a former passage existed to the open sea, however small. A third coastal category is intermediate between the others. Stretches of coastline are broken at intervals by former embayments that penetrated inland by varying amounts. Sedimentary rocks are more likely to be trapped and preserved within the bays, which represent more protected conditions than on outer rocky shores. Like deposits skirting island rocky shores, sedimentary layers in former bays often retain an original dip of about 6°. In this case, however, beds dip convergently like a funnel from the margins of the surrounding bay.

Examples of Pleistocene rocky shores also include small islands, linear segments on exposed outer shores, and semiprotected embayments. In most instances, these rocky shores qualify as wave-cut terraces (Ortlieb 1991) that were eroded from older substrates consisting of Cretaceous granodiorite, Miocene andesite, or Pliocene limestone. Marine deposits, such as coral reefs, occur intact on such terraces elevated from 12 to 70 m above the present sea level. The terraces end at landward walls that represent former sea cliffs. Particularly where sea cliffs were cut in Pliocene limestone, borings made by Pleistocene pholadid bivalves remain as evidence that seawater splashed against former shores. In other places, the fossil remains of red coralline algae, barnacles, and oysters are heavily encrusted on former sea cliffs composed of igneous rocks.

Studying Abandoned Rocky Shores

What we know about abandoned rocky shores is the result of more than a year of field studies on the Baja California peninsula and related islands, but extended over 20 years of research with the field component usually limited to the month of January. Geological maps compiled by McFall (1968) for the region surrounding Bahía Concepción in Baja California Sur and by Gastil et al. (1973) for the entire northern state of Baja California played a crucial role in guiding us to areas where we expected the landscape to reveal its paleogeography. The desire to see what lay over the next hill or around the next bend in the coast also led to fortuitous discoveries in unmapped parts where few if any geologists had gone before.

In the late 1990s, we began to realize that satellite images were helpful in finding new study sites and that former shorelines might be traced with greater accuracy based on the reflectivity of limestone against darker, adjacent rocks. By then, we had expended much time and energy walking the backcountry around the base of the Concepción Peninsula. Although of poor quality, the first satellite image we examined for this area was clear enough to give another perspective on the region and inspired new explorations in adjacent areas around San Nicolás to the southeast (fig. 3.1A). When an effort was launched to visit some of the islands in the lower Gulf of California in January 2003, satellite images proved to be indispensable. Observations by Anderson (1950) and Durham (1950) on islands such as Carmen, Monserrat, and San José were valuable, but limited in scope. Likewise, the geological map by Gastil et al. (1973) for Isla Angel de la Guarda off Bahía de Los Angeles provided the motivation to go there. Although powerful as a research tool, satellite imagery is no substitute for on-the-ground exploration. Fieldwork remains necessary to determine what kind of limestone and conglomerate deposits are arrayed on different kinds of rocky shores.

Pliocene Island Rocky Shores
Coyote Mountain Archipelago

Laguna Salada on the Mexican side of the international border and the Salton Sea in southern California show that a 250-km extension of the Gulf of California northwest from the present delta of the Colorado River fits with the regional geography. About 25 km north of the international border, the abundant marine fossils in the Imperial Formation in the Coyote Mountains (fig. 3.1A, locality 1) demonstrate that the gulf reached farther north during the Early Pliocene (Carreño and Smith 2007). Former islets are represented by small inliers of Paleozoic marble and other metamorphic rocks surrounded by Pliocene conglomerate and limestone. Some inliers express nearly vertical walls against which abutment unconformities with Pliocene conglomerate formed. Watkins (1990) mapped about 100 m of rocky shoreline around parts of an inlier, showing a range of trace fossils made by Pliocene animals on marble sea cliffs. They include club-shaped borings (*Gastrochaenolites*) made by

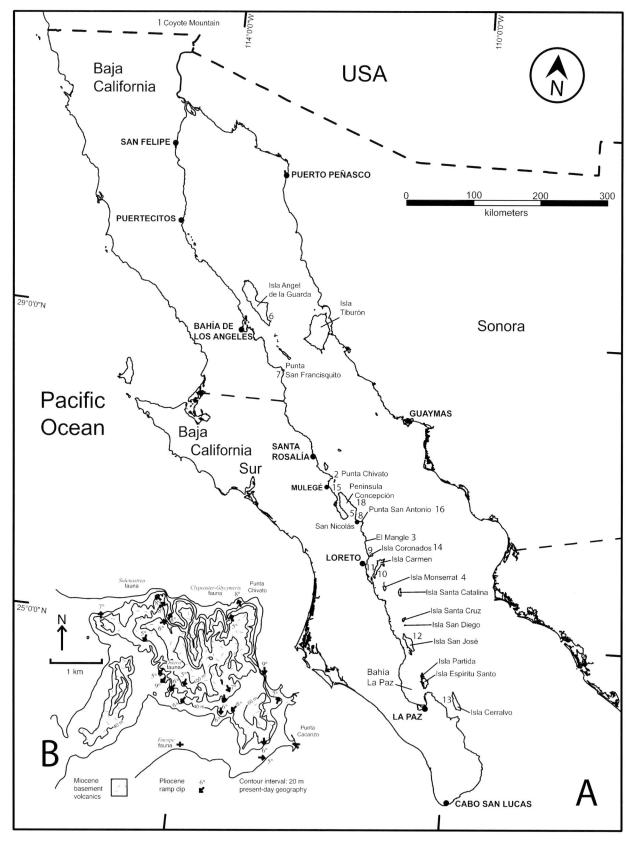

Figure 3.1. (A) Map of the Gulf of California with numbered localities marking locations for all Pliocene basins and Pleistocene marine terraces discussed in the text as areas with abandoned rocky shores; and (B) a detailed map of the Punta Chivato area.

pholadid bivalves, box-work borings (*Entobia*) made by clionid sponges, and thin borings (*Trypanites*) made by polychaete and/or sipunculate worms. The highest density of bivalve borings in marble is about six to eight over 10 cm^2 based on outcrop photos (Watkins 1990, p. 169). Abutting conglomerate includes pebble- to boulder-sized clasts of marble, schist, or quartzite. Marble boulders up to 90 cm in diameter typically are covered by bivalve borings. Subhorizontal unconformities in the Coyote Mountains also exhibit the colonization of corals (*Siderastrea californica*) and associated oysters directly on schist bedrock (Watkins 1990). These and other corals such as *Solenastrea fairbanksi* typically are restricted to a position in the Lower Pliocene of the Imperial Formation (Foster 1979). Little of the Coyote Mountains has been surveyed for paleoislands, but the part studied in detail by Watkins (1990, fig. 1) suggests the occurrence of individual islets up to 2 km in circumference.

Santa Inés Archipelago

Lower Pliocene strata at Punta Chivato between Santa Rosalía and Mulegé (fig. 3.1A, locality 2) are preserved intact as limestone ramps that encircle several small paleoislands named by Simian and Johnson (1997) as the Santa Inés Archipelago. A detailed series of paleogeographic maps was compiled to show development of the islands from fault blocks of Miocene andesite in the Comondú Group (Johnson 2002b). On average, the overlying limestone ramps exhibit 6° dip in a radial pattern around the fault blocks (fig. 3.1B). They record a prolonged relative rise in sea level broken by a short regression. At the start of the transgression, the largest of the paleoislands was embedded in the high ground of Mesa El Coloradito and Mesa Atravesada, the latter terminating to the east at Punta Chivato. The maximum circumference of the paleoisland conjoining the two mesas was about 13 km. As the block foundered under rising sea level, variations in elevation led to a cluster of four, smaller paleoislands. West of Punta Chivato at Mesa Ensenada de Muerte, a separate fault block with a circumference of 7 km was transformed into another paleoisland. A third fault block southeast of Punta Chivato with a minimum circumference of about 5.5 km is the foundation for three small islands in the present Gulf of California. The modern islands, one of which features a Pliocene limestone ramp, lend their name to the greater Pliocene archipelago of the Punta Chivato region.

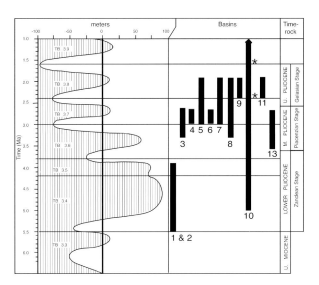

Figure 3.2. Standard Pliocene chronostratigraphy integrated with the eustatic sea-level curve of Haq et al. (1987) to show correlation of a dozen different basins in the Gulf of California. Vertical bars represent the age span (Ma) of basins (numbered): (1, 2) Coyote Mountain and Isla Santa Inés archipelagos; (3) El Mangle; (4) Isla Monserrat; (5) Santa Rosaliíta on Peninsula Concepción; (6) Isla Angel de la Guarda, southern end; (7) San Francisquito; (8) San Nicolás; (9) Punta El Bajo; (10) Arroyo Blanco; (11) Bahía Marquer; and (13) Isla Cerralvo. Stars adjacent to column 10 mark levels with red-bed deposits interpreted as small landward lagoons in the Arroyo Blanco section. The same column numbers apply to all basin locations shown in figure 3.1A.

A basal conglomerate with eroded andesite cobbles and boulders starts the ramp succession that blankets the Punta Chivato fault blocks. The limestone matrix of the conglomerate includes fossil corals, echinoids, and diverse mollusks attributed to the San Marcos Formation (Simian and Johnson 1997). The coral *Solenastrea fairbanksi*, bivalve *Glycymeris maculata*, and heavy echinoid *Clypeaster bowersi* are restricted to the northern side of the largest paleoisland (fig. 3.1B), where the thickness of the conglomerate and size of reworked andesite cobbles and boulders indicate high energy from wave activity. Oyster beds with large individuals that reached a shell length of 25 to 30 cm are restricted to limestone ramps on the southern side of the main paleoisland, where they are associated with a thin wedge of conglomerate containing smaller clasts (fig. 3.1B). A more protected environment with reduced wave activity is interpreted as having existed on the southern flank of the paleoisland (Simian and

Johnson 1997; Johnson and Ledesma-Vázquez 2001). The time frame for the Punta Chivato ramp succession is correlated with the basal part of a Pliocene sea-level curve (fig. 3.2, no. 2), meaning the initial transgression occurred at the beginning of the Pliocene.

Cerro Mencenares and El Mangle Fault Block

An extinct stratovolcano that covers 150 km² and rises to an elevation of 790 m is centered 30 km north of Loreto (fig. 3.1A, locality 3). Zanchi (1994) mapped primary faults associated with this feature, called the Cerro Mencenares volcanic center. Bigioggero et al. (1995) also described the igneous petrology and stratigraphy of flows within the complex, showing that volcanic activity occurred through parts of the Pliocene and Pleistocene. Our interest is with the sandstone and limestone beds that skirt much of the volcano's perimeter on its southern, eastern, and northern sides. It is possible that the entire complex represents one large paleoisland, but undisputed marine strata are difficult to trace on the western side.

Pliocene strata at El Mangle recline against the stratovolcano on the southeastern side of the complex. There, a ramp with a maximum thickness of 12.75 m dips 6° off an unconformity surface eroded in andesite to form a nearly continuous 2.5-km-long shoreline hugging the 40-m contour of the volcano (Johnson et al. 2003). The strata make a transgressive-regressive package with thick limestone beds at its core dominated by fossil mollusks. The paleoshore features onshore facies with abundant jackknife clams (*Tagelus californianus*), ark shells (*Anadara multicostata*), and strombid gastropods (*Strombus subgracilior* and *S. granulatus*). More sheltered parts of the paleoshore retain clusters of the coral *Porites panamensis* preserved in growth position and large oysters attached to andesite boulders. Down-ramp 300 m offshore is a distal facies dominated by fossil pectens, most of which are *Argopecten abietis* with minor representations of *Nodipecten subnodosus*. Barnacles encrust some of the fossil pectens. Fossil oysters (*Myrakeena angelica*) are present, but rare. Fossil echinoderms are less abundant than mollusks, but the echinoid *Clypeaster marquerensis* and sand dollar *Encope shepherdi* are common. These indicate a later Pliocene age (Durham 1950), but the sequence also overlies a tuff bed dated at 3.3 ± 0.5 Ma (Johnson et al. 2003). Thus, the time frame for the ramp succession at El Mangle is correlated with the third global cycle in the Pliocene sea-level curve (fig. 3.2, no. 3), meaning the peak

transgression took place during the transition from the Middle to Late Pliocene.

Pliocene limestone also flanks the northern side of Cerro Mencenares near Punta Mencenares 7 km north of El Mangle (fig. 3.1A). With an overall 6° dip to the north, these strata include distal facies with ark shells (*A. multicostata*) and abundant pectens (*A. abietis*). The heavy echinoid *C. bowersi* occurs in limestone from a more proximal or upslope position near the 100-m elevation, and this species is known to be restricted to Lower and Middle Pliocene strata (Durham 1950). On the opposite, southern flank of Cerro Mencenares, massive sandstone occurs about 5 km inland only a short distance south of El Mangle (fig. 3.1A). Laminations that resemble red coralline algae thickly encrust andesite cobbles from the basal conglomerate above the unconformity on the south flank of Cerro Mencenares. Otherwise, marine fossils are notably absent from this sector. Some wave activity was necessary to erode the cobbles, but the laminar rinds cover only the upper surfaces of the cobbles. By all indications, the stratovolcano's southern shore was a protected setting with poor marine circulation.

El Mangle Block is a fault-bound, 7-km-long paleoisland that formerly stood off the eastern coast of the Cerro Mencenares volcanic complex and developed a seaward ramp on which large populations of pectens thrived prior to transtensional faulting and uplift (see chaps. 9 and 12).

Isla Monserrat as a Pliocene Island

A small island (6.5 × 3.25 km), Isla Monserrat sits 45 km southeast of Loreto and 13 km off the closest shore on the Baja California peninsula (fig. 3.1A, locality 4). Spending only a day there as a member of the 1940 E. W. Scripps expedition, Anderson (1950, pp. 22–24) later described Pliocene limestone (up to 30 m thick) that sits with "marked angular discordance" on older Miocene volcanics. He also reported that basal conglomerate (variable in thickness but rarely exceeding 3 m) occurs between the limestone and underlying volcanics. An upslope view of beige limestone on tilted beds of red andesite near the southern end of Isla Monserrat strikes the classic pose of an angular unconformity true to Anderson's depiction. In this case, the limestone has the profile of a wedge that thickens from 3.5 to 10 m (fig. 3.3, left to right) on an erosion surface dipping from 6° to 10° to the southeast. The basal conglomerate and dip angles of overlying limestone beds found at many places on Isla Monserrat

Figure 3.3. (top) View looking west on the angular unconformity between Pliocene limestone with basal conglomerate and tilted Miocene andesite flows exposed in a fault scarp on the southern end of Isla Monserrat. Photo credit: Markes E. Johnson. Figure 3.4. (bottom) The highest elevation with a Pliocene paleoshore anywhere in the Gulf of California occurs 204 m above sea level on Isla Monserrat. Fossils from the paleoshore include the gastropods *Strombus galeatus* (large), *Nerita scabricosta*, and *Conus* cf. *princeps* (small); the bivalve *Cardita megastrophia*; and several corals. Photo credit: Markes E. Johnson

are consistent with interpretation of the ramp concept at Punta Chivato (Simian and Johnson 1997; Johnson 2002b) and El Mangle (Johnson et al. 2003), but the island is unique in the Gulf of California for the record-high elevations to which ramp deposits have been lifted.

On approach from the sea, the most impressive aspect of Isla Monserrat is the range of heights occupied by cap rocks of Pliocene limestone between sea level and almost the top of the island. The rocks are cut by a series of mainly north–south faults that give the

impression of giant steps ascending the island. Not all the Pliocene limestone on Isla Monserrat belongs to a ramp deposit (see chap. 9), but much can be correlated to a single coextensive ramp. The highest ramp elevation and related paleoshoreline occurs 204 m above sea level. A fauna that includes the gastropods *Strombus galeatus*, *Nerita scabricosta*, and *Conus* cf. *princeps*, the bivalve *Cardita megastropha*, and some corals (fig. 3.4) was found adjacent to a 50-m-long paleoshore in the south of the island. This is the proximal facies from the most elevated Pliocene ramp deposit we are aware of in Baja California. It means that only the southern end of the island once projected above sea level, with an area less than 5 percent the size of Monserrat today. An apt comparison is drawn with the tiny Islas Las Galeras, which sit 2 km north of Isla Monserrat but are surrounded by a broad, shallow-water shelf like a hat brim. Thrust the shelf 175 m upward and another Isla Monserrat is born.

Distal from the high paleoshore elsewhere on the ramp, *Argopecten antonitaensis* is an abundant component of the limestone and it signifies a Middle Pliocene position (Durham 1950). Due to faulting, the limestone surface there has an elevation of 175 m. The fault scarp exposes underlying conglomerate, which is 2.25 m thick and composed of eroded andesite boulders up to 36 × 40 cm in profile. The diverse fossil fauna belonging to the conglomerate includes several bivalves (*A. multicostata*, *Codakia distinguenda*, *G. maculata*, *Trachycardium procerum*, *A. abietis*, and *Lyropecten subnodosus*), gastropods (*S. galeatus*, *Fasciolaria princeps*, *Conus pluncticulatus*, and *Olivella tergina*) and two echinoids (*Clypeaster bowersi* and *C. revellei*). Overlap in biostratigraphic range of the two echinoids confirms a Middle Pliocene age for the conglomerate (Durham 1950). The Monserrat ramp is only slightly older than the ramp at El Mangle, but much younger than at Punta Chivato. Thus, it appears to represent the initial rise in sea level correlated with the third Pliocene highstand in sea level (fig. 3.2, no. 4).

Pliocene Rocky Shores on Lagoons
Santa Rosaliíta Basins on Peninsula Concepción

Middle to Upper Pliocene strata fill several shallow basins at the base of Peninsula Concepción near Rancho Santa Rosaliíta (fig. 3.1A, locality 5). A landscape 63 km² in area was mapped and interpreted with geological cross sections (Johnson et al. 1997; Johnson and Ledesma-Vázquez 2001), showing that limestone- and chert-replaced strata from the Infierno Formation make abutment unconformities against surrounding andesite hills of Miocene age. Central as a landmark, the 1-km² Cerro Prieto rises as a distinct paleoisland 150 m above a maze of small, lagoon-like basins. Oysters with an elongated shell (*Ostrea californica osunai*; Hertlein 1966) crowd inner-lagoon limestone deposits. Some oysters are quite large, up to 45 cm in length. All basins are interconnected by a series of restricted passages that lead to Bahía Concepción on the west but fail to cross to the open Gulf of California on the east. Different basin floors occupy different elevations, and a series of three paleogeographic maps shows the gradual flooding of the entire area (Johnson et al. 1997). A more diverse invertebrate fauna occurs in limestone from the passages and basin adjoining Bahía Concepción. Fossil barnacles encrust sea cliffs. The fossil bivalves *Dosinia*

ponderosa and *T. californianus*, together with the gastropod *S. subgracilior*, are abundant. Less common is the echinoid *C. marquerensis*, which implies a Late Pliocene age (Durham 1950). The stratigraphy of the Infierno Formation entails at least one major regression between two distinct highstands in sea level (fig. 3.2, no. 5). During maximum flooding (Johnson et al. 1997, fig. 11), a 35-km-long rocky shore followed twists and turns through a landscape dominated by andesite hills.

Pond Basins on Isla Angel de la Guarda

Pliocene conglomerate and siltstone are deposited over the southeastern part of Isla Angel de la Guarda, where inliers of andesite or basalt rise as monadnocks or small paleoislands through the surrounding strata (Gastil et al. 1973). Anderson spent two days on the

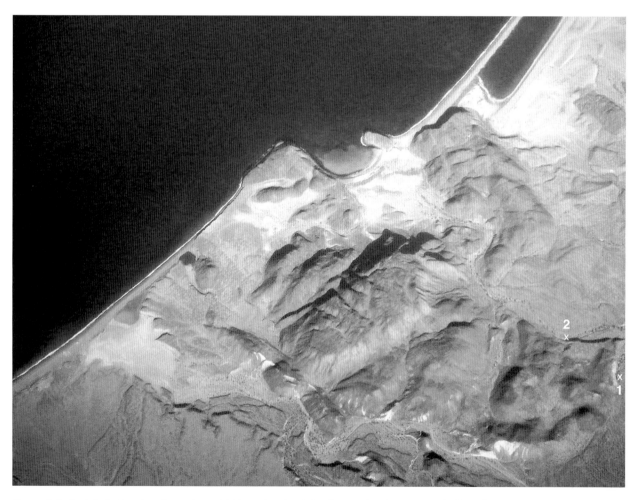

Figure 3.5. Aerial photo taken from about 9000 m showing island monadnocks surrounded by Pliocene strata on Isla Angel de la Guarda. Pliocene oysters from the subspecies *Ostrea californica osunai* occur well inland from the present coast (localities 1 and 2). Photo credit: Huajie Cao

island as a member of the 1940 Scripps expedition and later described gravel-rich strata from the southern end with a maximum thickness of 45 m (Anderson 1950, pp. 38–41). A modern, closed lagoon at the southern end is bordered on the landward side by 10-m-thick Pliocene siltstone with intercalated thin conglomerate beds. The conglomerate includes scattered oysters and rare sand dollars belonging to *Encope angelensis*. The latter fossil was identified by Durham (1950, p. 44) from the cliffs at this locality and assigned a Middle Pliocene age. The strata suggest that a restricted lagoon occupied the same spot for a long time.

Pond-like Pliocene deposits on southern Angel de la Guarda cover more than 20 km² flanked by andesite highlands to the southwest. The 45-m-thick conglomerate described by Anderson (1950) is a proximal deposit derived from these highlands. Flanking Pliocene deposits on the north are four low andesite inliers that allowed seawater to flow through narrow passages from the open gulf to the inner lagoons. Two large paleoislands with elevations that rise 50 m above the surrounding landscape are separated by a canyon that leads to a secluded lagoon (fig. 3.5). Rare oysters and pectens are found at widely separated spots (fig. 3.5, localities 1 and 2), where horizontal siltstone beds up to 14 m thick abut against andesite shores. We attribute fossil oysters from the inner lagoons to *O. californica osunai*, the same subspecies identified by Hertlein (1966) on Peninsula Concepción. Other fossils from this part of Angel de la Guarda include *Ostrea vespertina* and the pectens *Argopecten circularis* and *A. abietis* (Durham 1950, p. 21). Based on overall paleogeographic relationships, rocky shores around sheltered lagoons stretched 25 km in length during a Middle Pliocene rise in sea level (fig. 3.2, no. 6).

Pliocene Rocky Shores on Embayments
San Francisquito Embayment

Pliocene strata occupy 10 km² in an embayment on the peninsular coast at San Francisquito about 75 km south of Isla Angel de la Guarda (fig. 3.1A, locality 7). The paleogeographic map for this area (fig. 3.6) shows access by a narrow channel from the north and a broad opening from the east. Cretaceous granodiorite surrounds the basin on three and a half sides, making this one of the best examples to show the influence of granitic shores on limestone strata in a confined basin. For this reason, the satellite image chosen for this chapter covers the region around San Francisquito

(fig. 3.7). Large-scale relationships entrained in this landscape offer the opportunity to examine biofacies patterns and depositional processes on rocky shores both outside and within the Pliocene basin.

Only 400 m long, the northern channel at San Francisquito is defined by a pair of faults that make a graben. Granodiorite hills rise on the western and eastern sides of the inlet to elevations between 100 and 175 m. Remnants of a 5-m-thick ramp that slopes 6° to the north appear on the inner western side of the channel at an elevation 48 m above sea level. The ramp sits on granodiorite and it is assumed that a matching feature once existed on the opposite shore. Eroded igneous cobbles up to 8 × 11 cm in outline form a 1.5-m-thick conglomerate at the base of the ramp. Red coralline algae encrust cobbles, and fossil barnacles are attached to the algae. Spherical pebbles 3 cm across are fully encrusted by calcareous rinds up to 1.5 cm thick. These qualify as rhodoliths, and frequent agitation was required to enclose the clasts with concentric laminations of algal growth. *Nodipecten arthriticus* and *Pecten vogdesi* commonly occur in the overlying limestone, also encrusted by barnacles. Among other fossils is the echinoid *C. revellei*, which is diagnostic for the Middle Pliocene (Durham 1950). Overall, a high-energy setting in shallow to intertidal water is interpreted outside the main basin.

Another entrance to the Pliocene basin crosses the 2.5-km-wide beach at Ensenada Blanca, anchored on the north and south by granodiorite hills between 100 and 165 m in elevation (fig. 3.6). Inside the basin, the rocky shoreline traced by the Cretaceous-Pliocene nonconformity follows an arc 14 km in length. Elevations at which the nonconformity is exposed vary due to fault relationships. Two boundary faults are associated with the basin (Gastil et al. 1973). One extends east–west along the northern margin of the basin. The other runs north–south along the inner eastern margin, but also cuts through limestone and granodiorite to reach the outer coast. These and related normal faults buried below the basin facilitated erosion of a topographic depression prior to initial flooding during the Middle Pliocene. Moreover, the north-channel graben follows lineaments parallel to transform fractures still active in the Gulf of California. Faults on the graben show projections that intersect both entrances to the basin (fig. 3.6), which implies that transtensional tectonics prepared the way for flooding.

Boundary faults inside the basin were active during flooding, but also remained active for some time afterward. As evidence, Pliocene limestone dips away

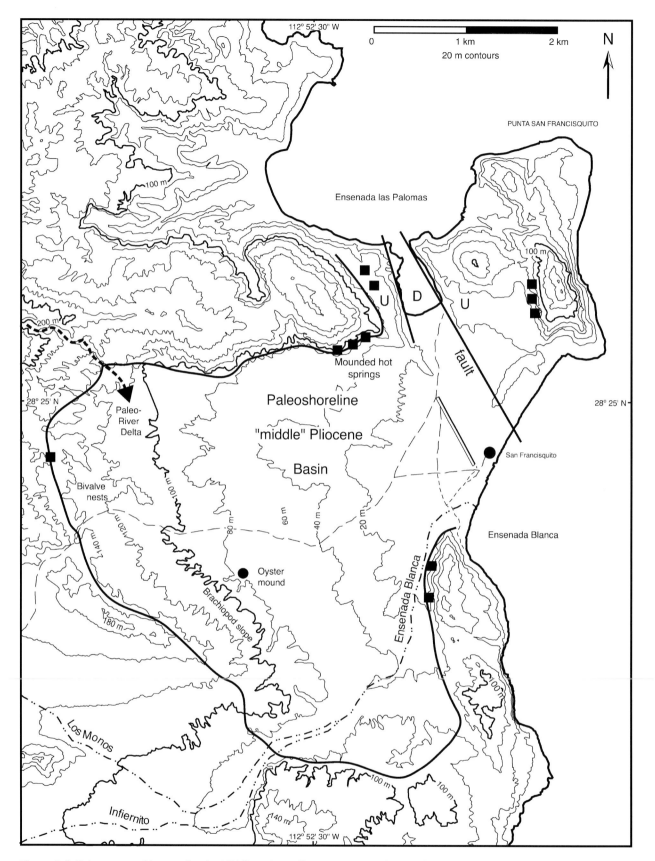

Figure 3.6. Paleogeographic map for the Middle to Late Pliocene San Francisquito basin showing basin dimensions with rocky shorelines widely emplaced at about 140-m elevation above present sea level.

Figure 3.7. False-color satellite image for (A) the coastal region from Punta Ballena west (including the Pliocene Ballena basin) and (B) from Punta San Francisquito south (including the Pliocene San Francisquito basin). The field of view for this image is approximately 24 km (east–west) x 18 km (north–south). In this band combination (7,4,1), limestone deposits show up in the San Francisquito basin as tan to greenish brown patches adjacent to the maroon coloration of Cretaceous granodiorite. Image acquired from University of Maryland Global Land Coverage Facility. The original data set was collected by the NASA LANDSAT Program, LANDSAT ETM+ scene L7CPF20000719_20000930_10, L1G, USGS, Sioux Falls, September 3, 2003.

from the northern paleoshore at inclinations internally adjusted between 20° and 25°. High dip angles suggest that the ramp was steepened after deposition, but syndepositional steepening also occurred during ramp accretion on the basis of changing dip angles. In terms of composition, the northern ramp is a silty limestone with abundant sea urchin spines and some disarticulated oysters as primary fossils. Layers within the northern ramp, which now intersects granodiorite at an elevation of 95 m, include only scattered granitic pebbles. Interpretation of the northern and eastern shores as parts of the embayment most sheltered from waves is supported by similar ramp lithologies with high silt content and low granitic detritus. Additional evidence for reactivation comes from localized hydrothermal deposits near the boundary faults (fig. 3.6).

Pliocene limestone dips from 6° to 10° toward the basin center from an elevation of 140 m on the western and southern sides of the embayment. Granitic

cobbles and boulders with a maximum diameter of 70 cm accumulated as conglomerate along the western paleoshore. Decrease in clast size toward the top of the conglomerate reflects a drop in wave energy with rising water depth. Normally, the basal conglomerate is less than 2 m thick, but reaches a maximum thickness of 8.5 m where a former stream brought large amounts of eroded granite to the northwestern corner of the bay. Populations of the bivalve *G. maculata* commonly are nestled in growth position among cobbles and boulders. The same bivalve is associated with mounded hydrothermal vents located near the northwestern paleoshore (see chap. 12). Besides the dominant *G. maculata*, other bivalves from overlying limestone include *A. multicostata, Cardita affinis,* and *C. megastropha.* The stratigraphic range for these bivalves extends back through the Pleistocene to the Late Pliocene (Durham 1950). The basal conglomerate on the basin's southern paleoshore typically incorporates articulated shells of *Ostrea fisheri*. Extensive brachiopod populations belonging to *Laqueus erythraeus* occur in limestone on the southern ramp, some preserved in crowded growth position above the 100-m elevation.

A somewhat older age is indicated by index fossils from the outer ramp of the northern channel than for the inner western shore of the embayment, but a marine incursion correlated with the transition from Middle to Upper Pliocene strata is reasonable (fig. 3.2, no. 7).

San Nicolás Embayment

As defined by Ledesma-Vázquez et al. (2006), the mixed succession of limestone, sandstone, conglomerate, and volcanoclastic rocks that compose the 150-m-thick San Nicolás Formation was deposited over 80 km^2 in a north-facing basin on Bahía San Nicolás (fig. 3.1A, locality 8). The basal unit is the Toba San Antonio tuff, which drapes a fault block in the southeastern corner of the basin. A radiometric age of 3.3 Ma for this tuff establishes the starting point for the Pliocene sequence. The basin is box-shaped, bound by strike-slip faults on the eastern and western sides and a normal fault on the southern side. Prior to incursion of marine water, a significant part of the San Nicolás Formation was emplaced as alluvial fan deposits that carried coarse debris from fault scarps to depocenters barred on the north by tilted fault blocks. Because the eastern boundary fault at Punta San Antonio brought Cretaceous granodiorite to the surface, contiguous fan deposits include granitic and andesitic materials. Granitic rocky shores, however, were minor in extent compared to the andesite footwalls on the faulted perimeter of the basin with a linear extent of about 30 km.

Excluding thick biocalcarenite deposits that show characteristics of strong tidal influence in the generation of sand waves with steep inclinations from 12° to 19°, other limestone units conform to the ramp model with whole fossils preserved in beds that dip 6° north toward the open gulf. The highest point in the San Nicolás Embayment, where Pliocene ramp limestone sits directly on Miocene andesite, is at about the 100-m elevation. An exposure of a prominent unconformity near the western margin of the basin was photographed by Johnson and Ledesma-Vázquez (2001, p. 74), illustrating a rare accumulation of fossil pen shells (*Pinna* sp.) that were transported from

a semi-infaunal setting to a final resting place within a large cavity on a rocky shore. Fossils from the San Nicolás Formation listed by Ledesma-Vázquez et al. (2006, table 1) show diverse mollusks, including *G. maculata*. Overlap of the heavy echinoid *C. bowersi* and pecten *Argopecten antonitaenis* suggest a Middle Pliocene age, but also the sand dollar *Encope* sp. and pecten *A. circularis* indicate a Late Pliocene age. Reinforced by the radiometric age calculated for the basal tuff within the basin, the longevity of the San Nicolás Embayment is regarded as Middle to Late Pliocene (fig. 3.2, no. 8).

Loreto Embayment

Continuous exposures of tilted Pliocene strata follow for as much as 5 km along the canyon walls of Arroyo Gua and Arroyo Arce near Loreto. Framed by Cretaceous granodiorite to the south and Miocene andesite to the north, Pliocene conglomerate, sandstone, and limestone cover more than 80 km^2 within the Loreto Embayment. This area has been exhaustively studied for its tectonics (Anderson 1950; Umhoefer et al. 1994), stratigraphy and biostratigraphy (McLean 1989; Dorsey et al. 1997; Carreño and Smith 2007), and paleontology (Durham 1950; Piazza and Robba 1998; Dorsey and Kidwell 1999). The tilted sedimentary layers in the Loreto Embayment do not fit the ramp model applied to Pliocene deposits elsewhere in the Gulf of California. In the southeastern part of the basin, Dorsey et al. (1997) recognized primary dips from 20° to 30° related to large-scale foreset beds and the kind of distinctive contacts produced between topset and foreset beds in fan-delta deposits. Sheltered rocky shores within the basin are poorly exposed along the northern abutment unconformity between Pliocene conglomerate and Miocene andesite, but entail a map distance of about 10 km. With regard to paleoecology, small oyster mounds sit directly on Miocene andesite in some places where the unconformity is close to horizontal (Dorsey et al. 1997, p. 90).

Outside the Loreto Embayment near Punta El Bajo (fig. 3.1A, locality 9), an unconformity and related Upper Pliocene ramp complex occupy an exposed rocky-shore setting. Miocene andesite with topographic relief upward of 600 m was the backdrop for recurrent ramp couplets, each consisting of a boulder conglomerate paired with overlying limestone rich in rhodolith debris (Dorsey 1997). Few megafossils occur through the ramp succession, but echinoids such as *E.*

shepherdi suggest a Late Pliocene age (Durham 1950). With a westward dip of 24° toward the open gulf, the innermost ramp is steepest. The next ramp dips 18° in the same direction, overlapped by a third that dips 15°. With a dip of only 11°, the outermost ramp exhibits the least incline. Dorsey (1997) proposed episodes of fault-controlled uplift along a tectonically active paleoshore to account for the couplets, each with a different dip angle. During deposition, each ramp probably accrued on the same natural slope of 6°. As the ramps became shingled on top of each other, new episodes of uplift had the cumulative effect of tilting older ramps more than the younger ramps. The entire complex correlates with the fourth and final rise in global sea level during the Late Pliocene (fig. 3.2, no. 9).

Pliocene Embayments on Large Islands

Carmen, San José, and Cerralvo islands all retain Pliocene embayments enclosed by former rocky shores (fig. 3.1A, localities 10–13). Each island shows a complicated tectonic relationship to the surrounding seafloor on the Gulf of California. While geological signatures can be related to the peninsular mainland to show some amount of lateral movement, the same islands existed as large paleoislands through much of the Pliocene. In other respects, however, the island basins are similar in size and orientation to those on the nearby Baja California peninsula.

The Arroyo Blanco and Bahía Marquer basins partially overlap in age and are situated on the opposite flanks of Isla Carmen. With a smaller area of 3.87 km², the Arroyo Blanco basin faces the open gulf on the eastern side of the island (fig. 3.1A, locality 10). The northern and southern margins are defined by divergent normal faults that give the basin a distinctly trapezoidal outline. No obvious fault boundary occurs at the rear of the basin, where marine Pleistocene strata pinch out against Miocene andesite at the 170-m elevation to make a low-relief rocky shore. Based on continuous canyon exposures through the deep drainage system of Arroyo Blanco, Eros et al. (2006) described a 157-m-thick sequence of Pliocene-Pleistocene strata within the basin. Although the canyon fails to reveal the contact between Pliocene strata and underlying andesite, Pliocene strata account for 90 percent of exposed sedimentary rocks. Ramp inclinations between 4° and 6° are maintained through the Pliocene sequence, which includes a huge volume of

limestone composed of crushed rhodolith debris (see chap. 7), lesser shell beds dominated by pectens, and thin conglomerate beds composed of andesite cobbles and boulders.

Overlapping range zones for 22 fossil species tracked on the canyon walls of Arroyo Blanco (Eros et al. 2006, fig. 6) show that the basin underwent ramp accretion from the Early to Late Pliocene. *Argopecten sverdrupi* and *Lyropecten modulatus* are restricted to horizons low in the sequence, for example, while the echinoid *E. shepherdi* appears only high in the sequence. Thin lagoon deposits with terrigenous clay punctuate the marine Pliocene succession in at least two places. The Arroyo Blanco basin on Isla Carmen (fig. 3.2, no. 10) had a demonstrably longer history than any other Pliocene basin correlated in this study. Andesite rocky shores that surround the basin on three sides are about 5 km in length.

Best exposed in sea cliffs on the opposite side of the island, strata held by the Bahía Marquer basin cover 3.42 km² in area and dip gently outward toward the Carmen Passage south of Loreto (fig. 3.1A, locality 11). Anderson (1950) named the Marquer Formation for the fossil-rich strata deposited here. A strictly Upper Pliocene position is attributed to this formation (fig. 3.2, no. 11), based on the diverse fossil fauna studied by Durham (1950). Numerous northwest–southeast trending faults with small offsets dissect the basin, making it more difficult to reconstruct than the neighboring Arroyo Blanco basin. Gross paleogeographic features include a narrow peninsula formed by a low andesite ridge that projects from the rear of the basin at about the 80-m elevation. An effective bifurcation of the basin into northern and southern sectors is the result. Both sectors preserve thick wedges of andesite conglomerate that dip from 4° to 6° off landward margins to interfinger with distal limestone beds. A conglomerate with andesite pebbles forms an apron around the nose of the west-pointing peninsula, and small fossil oysters are preserved in growth position cemented to individual clasts. With its internal peninsula, the Marquer Embayment developed a rocky shoreline no less than 11 km in length. Up to 17 m in thickness, the distal ramp limestone exposed in sea cliffs on Bahía Marquer is highly enriched in crushed rhodolith debris.

Pliocene basins in the northeastern part of Isla San José open eastward onto the Gulf of California (fig. 3.1A, locality 12). Umhoefer et al. (2007) summarized the general stratigraphy and related fault boundaries associated with transtensional tectonics that define these basins. Sedimentary strata on Isla San José reach a cumulative thickness of more than 1,800 m, much of which is attributed to alluvial-fan deposits with coarser conglomeratic facies more proximal to fault scarps and pebbly sandstone in a more distal position (Umhoefer et al. 2007, table 1). Pliocene strata on San José mostly form nonconformities against Cretaceous granodiorite with angular unconformities on Miocene andesite a distinct second in frequency. Massive sandstone derived from eroded granitic shores accumulated in a deltaic setting, while bedded sandstone with a shell-hash component was deposited farther offshore. A more distal mudstone unit contains foraminifera that reflect ages between 2.4 and 2.0 Ma, while a tuffaceous sandstone unit yields a conservative estimate of 3.1 Ma (Umhoefer et al. 2007). Thus, the flooding of the basins occurred during and mostly after the change to transtensional tectonics in the region. Based on analysis of satellite images, the cumulative distance represented by granitic rocky shores around the San José basins is not less than 8 km. The age span of the San José basins is uncertain.

Hertlein (1966) recorded aspects of a small Pliocene basin on the western side of Isla Cerralvo facing peninsular Baja California across the Cerralvo Channel (fig. 3.1A, locality 13). Maximum stratigraphic thickness is less than 170 m, with a distinctive 10-m thick unit of calcarenite composed of crushed rhodolith debris (see chap. 7) preceded and followed by units with conglomerate and sandstone that account for most of the basin's volume. Postdepositional faulting has destroyed any semblance of an original ramp configuration, but the main north–south fault brings Pliocene strata into contact with Cretaceous granodiorite. The maximum length of a Pliocene granitic rocky shore on Isla Cerralvo is only 0.75 km. Key fossils from the upper sandstone beds include the heavy echinoid *C. bowersi* and the shells of *Argopecten revellei*, which overlap to indicate a Middle Pliocene age for the basin (fig. 3.2, no. 13).

Late Pleistocene Rocky Shores

Isla Coronados near Loreto (fig. 3.1A, locality 14) is an example of a small island (3 × 3.5 km) that existed as a Pleistocene island looking much as it does today. Built on andesite flows dated to 690 ± 50 Ka and 160 ± 20 Ka (Bigioggero et al. 1988), the 260-m-high cone of an extinct volcano dominates the island's profile. Extensive coral gardens that date from about 121 Ka filled much of a great lagoon covering 50 ha at an elevation now close to 12 m above sea level on the southern side of the island (Johnson et al. 2007). Dense clusters of coral colonies (*P. panamensis*) up to 20 cm in height are preserved upright in tiers attached to the slopes of andesite barriers that frame the outer margin of the great lagoon. The inner shore follows the lower slopes of the volcano and it is possible to follow a 1-km paleoshore where white carbonate sand rich in coarse rhodolith debris abuts dark andesite rocks. Satellite imagery for Isla Coronados is refined enough to reveal the rocky shoreline of a tiny inner lagoon approximately 100 m long by 30 m wide isolated from the great lagoon by a narrow arm of volcanic rocks.

Cerro El Sombrerito is a landmark situated in the estuary of the Mulegé River (fig. 3.1A, locality 15) that fits the ideal characterization of a Pleistocene wave-cut platform with associated terrace deposit. This feature (fig. 3.8) is a Pleistocene islet with a 12-m terrace incised on the sides of a gabbro plug from the eroded chamber of a former volcano. Corals collected from the terrace deposit yielded an age of 144 Ka and the amount of local uplift necessary to raise the terrace to its present elevation was calculated as 4 to 5 cm per 1,000 years (Ashby et al. 1987). A contemporary wave-cut platform is being eroded around the base of the plug, which is approximately 250 m in circumference.

Libbey and Johnson (1997) discovered a Pleistocene islet named Isla Fantasma situated 600 m inland from Playa La Palmita near Punta Chivato (fig. 3.1A, locality 2). With a circumference of almost 1 km, the paleoisland reflects a low andesite rocky shore at an elevation of 11 m above the present sea level. The northern coast was abundantly populated by bivalves (*C. distinguenda*, *Periglypta multicostata*, and the smaller *Chione californiensis*) all articulated in life position and wedged among andesite cobbles adjacent to the rocky coastline. Large conch shells (*S. galeatus*) are common near the southern coast, where crushed rhodolith debris abuts against andesite.

Johnson and Ledesma-Vázquez (1999) studied a boulder bed composed of mixed andesite and granodiorite boulders forming a sea cliff that tracks 400 m around the margins of a large Pleistocene embayment near Punta San Antonio (fig. 3.1A, locality 16). Now elevated 27.5 m above sea level, the cliff top rises 14 m above the seabed of the abandoned bay. In addition to encrustations on boulders by coralline red algae, 11 species dominated by bivalves are found attached to or wedged in life position among the boulders and another 18 species dominated by gastropods

Figure 3.8. View looking south on Cerro El Sombrerito near Mulegé, where a 12-m Pleistocene terrace is incised on the sides of a gabbric pluton. Photo credit: Markes E. Johnson

and other unattached invertebrates also occur at this locality (Johnson and Ledesma-Vázquez 1999, tables 1–3). Bivalves commonly preserved in life position on the outer, more exposed part of the rocky shoreline include *Modiolus capax, C. distinguenda,* and *P. multicostata.* The gastropod *Turbo fluctuosus* is especially plentiful. Coral colonies (*P. panamensis*) up to 18 cm in height occupy the most sheltered part of the inner bay.

The outer shores of Peninsula Concepción record exposures of a 13-m marine terrace that may be traced for 25 km along the andesitic coast between the village of San Nicolás and Punta Verde (fig. 3.1A). Cobbles and small boulders up to 30 cm in diameter, some of which are encrusted by the barnacle *Tetraclita stalactifera* (Johnson and Ledesma-Vázquez 2001) sit atop the 13-m terrace at Paredon Amarillo on the northeastern tip of the peninsula (fig. 3.1A). The most widely represented bivalve on the 13-m terrace anywhere on the outer shores of Peninsula Concepción is *C. distinguenda.*

Older Pleistocene terraces are recorded throughout much of the gulf coast along the Baja California peninsula (Ortlieb 1991), but the more elevated terraces retain fewer fossils, which are rarely preserved in growth position. Eros et al. (2006, p. 1159) reported on an older Pleistocene fauna from Arroyo Blanco

on Isla Carmen (fig. 3.1A, locality 10), which also includes articulated *C. distinguenda* and *C. californiensis* but at an elevation 170 m above the present sea level. This site represents a rare Pleistocene ramp rich in rhodoliths nucleated around andesite pebbles and developed against an andesite shore at the rear of the Arroyo Blanco basin. Discontinuous uplift of the basin resulted in the subsequent erosion of at least four marine terraces now found at elevations of 68, 58, 37, and 12 m above sea level. All were cut in Pliocene limestone, but only the lowest terrace retains traces of Pleistocene bivalve borings in Pliocene limestone. The highest well-defined terrace at Arroyo Blanco was tentatively correlated by Eros et al. (2006) to the 65-m terrace of Ortleib (1991) from Punta Perico on Isla Carmen, which is regarded as about 400 Ka in age. The maximum distance any of the limestone terraces at Arroyo Blanco may be followed is 1.5 km.

Summary of Major Relationships
Pliocene and Pleistocene Biogeography

Seasonal winter winds and wind-driven waves that so strongly impact north-facing rocky shores in the present Gulf of California (chap. 2) find clear manifestations in Pleistocene-Pliocene biogeography traceable back 5.5 Ma. Through the Pliocene, a faunal association that included species belonging to the heavy echinoid *Clypeaster* and sturdy bivalve *Glycymeris* persisted on rocky shores with a northern exposure. Clypeasteroid echinoids are noted for thick tests, strong connections between component plates, and internal struts that combine to make a robust frame capable of withstanding high-energy environments (Nebelsick and Kroh 2002). Glycymerid bivalves settled among shore boulders, where the wedging habit offered security against dislodgment by waves. The *Clypeaster-Glycymeris* association occurs on the north-facing paleoshores of Punta Chivato, Cerro Mencenares, and San Nicolás (fig. 3.1A, localities 2, 3, and 8), as well as east- or west-facing paleoshores at Isla Monserrat and San Francisquito (fig. 3.1, localities 4 and 7) that were influenced by wave refraction to some extent.

Leeward settings in the Gulf of California are south-facing rocky shores or sheltered inlets. Oysters used this habitat during Pliocene time, as exemplified at Punta Chivato (fig. 3.1B). The large subspecies *O. californica osunai* also colonized pond-like lagoons surrounded by hills on Peninsula Concepción and Isla Angel de la Guarda (fig. 3.1, localities 5 and 6). Variations in salinity may have been a deciding factor in these more restricted settings, as oyster shells are sufficiently robust to withstand wave activity.

Large Pliocene bays with rocky shores that opened east or west typically became repositories for huge volumes of crushed rhodolith debris transported to a final resting place by wave activity. The Arroyo Blanco and Bahía Marquer basins on Isla Carmen (fig. 3.1A, localities 10 and 11) typify this relationship. Other paleoshores with similar orientations that accumulated extensive rhodolith debris are found outside the Loreto embayment near Punta Bajo and on Isla Cerralvo (fig. 3.1A, localities 9 and 13). Pectens proliferated on east- to west-facing ramps off Pliocene rocky shores. Fossil pecten beds generally are interbedded with rhodolith debris in the Arroyo Blanco and Bahía Marquer basins of Isla Carmen. Extensive pecten beds also occur off east–west paleoshores at Punta Chivato and also Islas Monserrat and Cerralvo.

Pleistocene distributional patterns mimic those of the earlier Pliocene with some variations. Large bivalves such as *C. distinguenda* with a wedging habit and robust clinging gastropods such as *T. fluctuosus* were favored on northern exposures. Examples of the *Codakia-Turbo* association occur on north-facing shores at Punta Chivato and nearby Isla Fantasma (fig. 3.1A, locality 2), as well as on west-facing shores near Punta San Antonio (fig. 3.1A, locality 16) and east-facing shores all along the outer Peninsula Concepción (fig. 3.1A, locality 18). In contrast, branching coral colonies of *P. panamensis* typically sheltered on leeward, south-facing shores at Punta Chivato and several Pleistocene islands, including Coronados, Carmen, San Diego, and Cerralvo (Johnson et al. 2007). Oysters often were the first invertebrates to colonize rocky surfaces in Pleistocene bays, but they were easily surpassed by branching corals in area of occupied territory. Rhodolith debris continued to accumulate as a depositional product in east- and west-facing Pleistocene bays, as at Punta Bajo and the Arroyo Blanco basin on Isla Carmen (fig. 3.1A, localities 9 and 10).

Sea-Level Fluctuations and Regional Tectonics

Interplay of global changes in Pliocene sea level and regional tectonics in the Gulf of California is suggested by the sea-level curve modified from Haq et al. (1987) in juxtaposition with correlations of time-rock units from Pliocene basins in the Gulf of California (fig. 3.2). A more recent sea-level curve by Miller et al. (2005) was compiled on the basis of isotopic variations and is more complicated to relate to unconformity patterns around the Gulf of California. The length of vertical columns in figure 3.2 corresponds directly to the time span in millions of years over which different gulf basins accumulated strata. The actual starting point for initial flooding within the San Nicolás and El Mangle basins is grounded on radiometric dates for rocks at or near the base of the stratigraphic successions (fig. 3.2, nos. 3 and 8). In all other cases, however, the relative placement of the numbered columns is based on biostratigraphic determinations proposed by Durham (1950) using concurrent range zones mainly from fossil pectens and echinoids. It is important to note that some stratigraphic sequences demonstrate a history of relative sea-level change, as for example at Punta Chivato and Santa Rosaliíta on Peninsula Concepción (fig. 3.2, nos. 1 and 5) that may be aligned with the global curve subject to biostratigrahic control.

The results show the spread of fully marine waters between Isla Carmen and the Coyote Mountains of southern California linked to the initial rapid rise in global sea level during the Early Pliocene. Global sea level at about 4.5 Ma may have exceeded the present level by as much as 90 m (Haq et al. 1987). In so much as columns 3, 4, 6, and 13 are limited to a Middle Pliocene position (fig. 3.2), those districts must have been elevated above sea level or closed off from rising waters during the Early Pliocene. Because marine life reached the Coyote Mountains during early Pliocene time, it signifies only that districts such as El Mangle were bypassed at that time or that Lower Pliocene strata near El Mangle are now below the water level.

In some regions such as Isla Angel de la Guarda, Punta Chivato, and Santa Rosaliíta on Peninsula Concepción, there is little or no sign of postdepositional tilting related to tectonic influences. Sedimentary beds remain flat lying in lagoonal settings or retain evenly radial syndepositional dips of about 6° around island cores. Elsewhere, interference by local tectonics is clearly evident. Because the narrow time frame for deposition of Middle Pliocene strata on Isla Monserrat came and went without any subsequent addition of Upper Pliocene strata (fig. 3.2, locality 4), it appears that the entire island experienced sufficient vertical uplift prior to the Late Pliocene to put the area beyond reach of rising global sea level at the end of the epoch. In contrast, the Arroyo Blanco basin on nearby Isla Carmen sustained virtually continuous tectonic subsidence through most of the Pliocene into the Early Pleistocene (fig. 3.2, no. 10). Drops in global sea level may have exceeded the rate of ongoing local subsidence during intermittent deposition of lagoonal deposits derived from terrestrial sources (Eros et al. 2006). A global drop in sea level at about 2.4 Ma, for example, may have fallen below the present level by 75 m or more (Haq et al. 1987; Miller et al. 2005).

Deposition of Pliocene strata in the San Francisquito and San Nicolás basins (fig. 3.2, nos. 7 and 8) was accomplished during two consecutive pulses of rising global sea level, perhaps abetted by continuous subsidence of those basins but more so during the intervening global drop in sea level. The San Francisquito basin also shows internal evidence of influence by transtensional tectonics that altered steepness both during and after ramp formation. Likewise, the Punta El Bajo sequence studied by Dorsey (1997) provides striking evidence of episodic uplift that progressively altered ramp inclinations during a simultaneous global rise in Late Pleistocene sea level (fig. 3.2, no. 9). It is significant that most of the Pliocene basins considered in this study were flooded after the region-wide regime change to transtensional tectonics that occurred at about 3.5 Ma (fig. 3.2, nos. 3–8 and 13).

Ramp configurations occur throughout the Pliocene Gulf of California, but only rarely with respect to Pleistocene relationships. The prevalence of wave-cut platforms and marine terrace deposits in the Pleistocene indicates another significant change but one still affected by transtensional tectonics. In contrast to the steady subsidence of the Arroyo Blanco basin on Isla Carmen during which ramp inclinations varied minimally between 4° and 6° (Eros et al. 2006), development of Pleistocene marine terraces required intervals of tectonic uplift broken by intervals of tectonic calm during which coastal erosion made better progress in cutting subhorizontal platforms. Although there are local exceptions where the rate of uplift exceeded the regional Pleistocene pattern (Ortlieb 1991), marine terraces on the gulf coast of the Baja California peninsula tend to be laterally uniform.

Pliocene-Pleistocene Coastal Rocks Quantified

Coverage of Pliocene-Pleistocene rocky shores is not comprehensive in this review, but their distribution is broadly representative across the full length of the Gulf of California (fig. 3.1A). To conclude, it is appropriate to tally the composite length of abandoned coastline and show the breakdown in terms of different rock types compared to data generated on modern rocky shores. Every bit of Pliocene and Pleistocene rocky coastline enumerated here comes to a total of 198 km, which amounts to only 14 percent of the 1,414 km tabulated in chapter 2 for present-day rocky shores. Our sample from the geological past is small, indeed, when averaged over 5.5 Ma. Yet, statistics inform us that proportional relationships with respect to coastal rock types were not very different from today.

One difference is that none of the abandoned rocky shores described here involve much in the way of metamorphic rocks or younger volcanics (basalt). Normalizing for the absence of these rocks in the data set, the following comparisons may be made. Miocene andesite is the most pervasive rock type that formed Pliocene and Pleistocene rocky shores, with 172 km or 87 percent of the whole, compared to 715 km from the adjusted base of 1,230 km for modern rocky shores (58 percent). Cretaceous granodiorite makes a respectful showing with 24 km of Pliocene-Pleistocene rocky shores (12 percent of the whole), compared to 294 km from the adjusted base for modern rocky shores (24 percent). Least surprising, Pliocene limestone is marginally represented by only 0.0075 percent of the 198 km tracked in this study, whereas 18 percent of contemporary rocky shores in the gulf region are composed of limestone. Pliocene and younger limestone beds had to be deposited and lithified before they could be reconfigured as present-day rocky shores. Overall, our window on past rocky shores in the Gulf of California is small compared to what can be observed today. Nonetheless, these Pliocene-Pleistocene rocky shores provide great insights regarding patterns of species distribution and the variable influence of tectonic factors.

4 Corals and Coral-Reef Communities in the Gulf of California

Héctor Reyes-Bonilla and Ramón Andrés López-Pérez

Since the earliest recognition of *Porites* from Isla Carmen in the Gulf of California (Grewingk 1848), coral reef studies in the region have increased dramatically due to the description of new hermatypic and ahermatypic species (Durham 1947; Squires 1959); the publication of numerous geographic range extensions (Cairns 1991; Reyes-Bonilla 1993a); a research agenda in topics as varied as bioerosion, community structure and dynamics, biogeography, symbioses, bleaching, and El Niño conditions (Reyes-Bonilla 1993a, 1998; Reyes-Bonilla et al. 2002, 2005a; Iglesias-Prieto et al. 2004; LaJeunesse et al. 2007); and a multitude of publications on coral-reef-associated faunas (Solis-Marin et al. 1997; Brusca et al. 2005; Reyes-Bonilla et al. 2005b; and references therein). Indeed, because of the numerous publications and their scope, the Gulf of California is the best-known coral-reef area along the Mexican Pacific coast. In this chapter, we summarize published and unpublished information on corals and coral-reef communities in the Gulf of California, and we use numerical techniques to define large-scale spatial areas and their main trends in species composition, diversity, and abundance.

Coral Distribution

The Gulf of California contributed 36 percent of the entire Mexican Pacific coral records (*n* = 3423; Reyes-Bonilla et al. 2005a), distributed across 78 sites mainly located around islands in the Gulf of California and western gulf shores and scattered along the coasts of Sonora and Sinaloa (Medina-Rosas 2006) on mainland Mexico (fig. 4.1). Eighteen hermatypic coral species (*Pocillopora* spp. [five], *Psammocora* spp. [four], *Pavona* spp. [three], *Fungia* spp. [three], *Porites* spp. [two], and *Leptoseris* [one]) inhabit gulf waters from Punta Peñasco (31.29°N), Sonora, to Cabo San Lucas (22.87°N), Baja California Sur. Species distribution is highly heterogeneous and per-locality species

is highly skewed, resulting from the high number of localities with one species (e.g., *Porites panamensis*, 31 localities). Indeed, only seven of the 78 localities register 10–13 species: Isla Carmen (10 species), Cabo San Lucas and La Paz (11 species), Isla Cerralvo and Isla San José (12 species), and Cabo Pulmo and Isla Espíritu Santo (13 species).

Agglomerative unweighted pair-group average cluster analysis, multidimensional scaling, and similarity analysis were used to analyze the presence/absence data matrix. These analyses demonstrate that the Gulf of California has two characteristic coral assemblages in the northern and southern areas of the gulf. A depauperate southern-derived fauna is distributed from Loreto (25.50°N) to Cabo San Lucas (22.87°N; global $R = 0.634$, $P = 0.001$; fig. 4.2A, B). The

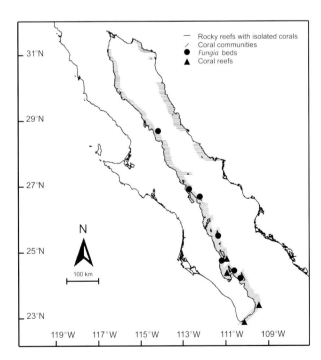

Figure 4.1. Map of the study area depicting the location of the main types of coral communities.

northern Gulf of California fauna ranges from Punta Peñasco (31.29°N), Sonora, to Punta Prieta (27°N), Baja California. Here, isolated monospecific coral patches are predominantly constructed by the encrusting phenotype of *P. panamensis*. However, south of Isla Tiburón (28°N), the massive and columnar phenotypes become common (Squires 1959), as developed on isolated rocks or patches of rocks surrounded by sand. In addition, occasional occurrences of *Porites sverdrupi* (Isla Angel de la Guarda), *Fungia curvata* (Bahía de Los Angeles), *Fungia distorta* (Isla Partida), and *Pocillopora verrucosa* and *P. capitata* (Isla San Marcos) do not alter the dominance of *P. panamensis* in the coral patches. The southern Gulf of California coral fauna ranges from Isla San José (25°N) to Cabo San Lucas (22.87°N), Baja California Sur. Here, development of reef structures occurs only at Cabo Pulmo (23.44°N) and to a lesser degree in Bahía San Gabriel and off Isla Espíritu Santo (24.42°N). In contrast, the other sites whose communities are part of the southern Gulf of California coral fauna (Isla San José, La Paz, Isla Cerralvo, and Cabo San Lucas) essentially are species-rich coral patches.

Southern gulf communities are monospecific, or nearly so, and consist of stands dominated by *Pocillopora* spp. Poritids and agariciids are relatively uncommon or restricted to deep waters and do not contribute much carbonate material to the reef structure. In particular, *P. verrucosa* is especially abundant (Reyes-Bonilla 2003). Finally, the southern-derived fauna extends from the Loreto area to the tip of the Baja California peninsula. In constructional terms, these are southern-derived, species-depauperate coral patches whose configuration is controlled by the arrangement of rock ridges upon which each patch is currently situated. Assemblage species composition shows no clear discontinuities between subgroups. Instead, similarity analyses demonstrate a close relationship (pairwise $R = 0.054$–0.333, $P = 0.13$–1.0) and a strong dependence on the southern Gulf of California coral fauna species pool (pairwise $R = 0.188$, $P = 99.6$).

Types of Coral Assemblages and Representative Faunas

The Gulf of California has four environments where reef corals occur: (1) isolated colonies or patches, (2) corals in rhodolith beds and other soft-bottom realms, (3) coral communities, and (4) actual coral reefs (fig. 4.1). The first category is dominant in Sinaloa, Sonora, Baja California, and the northern

part of Baja California Sur (26° to 28°N). In these areas, corals usually cover less than 1 percent of the bottom and build no framework due to their small size. At most, they are 30 cm in height or width. Instead, the colonies appear in an independent manner, add no significant substrate heterogeneity, and usually embody just one additional member of the diverse encrusting fauna from the central and northern gulf (Brusca 1980; Reyes-Bonilla et al. 2008). These habitats usually present only one or two reef coral species (*P. panamensis* and sometimes *P. sverdrupi*). Ecologically, there is little difference from a rocky reef in the sense that the primary producers are algae (turf and fleshy) and the system is relatively simple. The fauna is dominated numerically by herbivore invertebrates (mostly sea urchins of the genus *Arbacia* and starfishes *Pharia*, *Phataria*, and *Pentaceraster*) and supports high levels of richness and abundance in carnivorous fishes (Viesca-Lobatón et al. 2008).

Although corals are usually associated with hard-bottom areas, the Gulf of California nurtures unusual assemblages in sand and gravel areas adjacent to rocky coasts, as well as in rhodolith beds (see chap. 7). In these environments, corals of the genera *Porites*, *Psammocora*, and occasionally *Pavona* are not cemented to the bottom, but instead appear as coralliths. That means they function as free-living, rounded colonies, which when small can be moved by currents and suffer fragmentation (Reyes-Bonilla et al. 1997). After some growth in the sand, corals gain sufficient size to become fixed in place due to their weight or become anchored by their branches. Eventually, most colonies at this stage end up buried in the sediment. In addition to coralliths, Reyes-Bonilla et al. (1997) and Reyes-Bonilla (2003) found that large populations of mushroom corals (*Fungia* beds; figs. 4.1, 4.3A) exist in the gulf. Glynn and Wellington (1983) first described this habitat in the eastern Pacific in the Galápagos Islands. Three of these beds have been located, so far, at depths from 25 to 30 m between Bahía La Paz (24°N) and Isla Monserrat near Loreto (25°N). All are dominated by *F. distorta*, a self-fragmenting coral (Colley et al. 2002), with the occasional presence of *F. curvata*, *Psammocora stellata*, and very rarely *Pavona gigantea*. The conspicuous associated fauna is composed of sea urchins (especially *Toxopneustes roseus*), gastropods (*Strombus galeatus*, *Muricanthus* sp.), and bryozoans (James 2000; James et al. 2006).

The most common assemblage in shallow areas of the southwestern Gulf of California is the coral community, which typically can be observed from 22° to

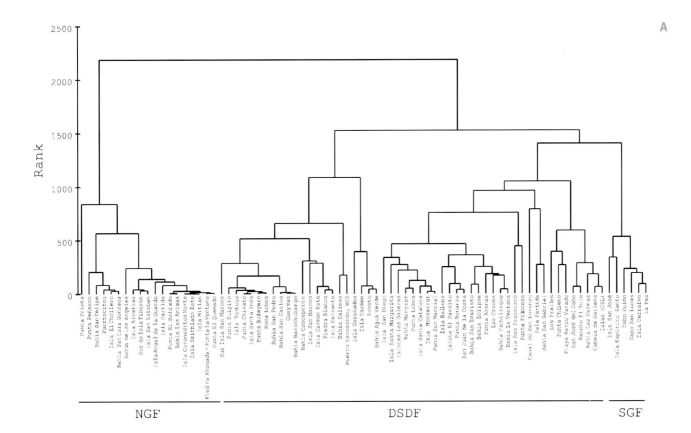

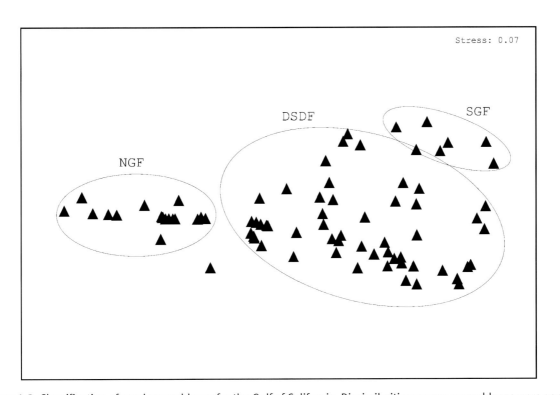

Figure 4.2. Classification of coral assemblages for the Gulf of California. Dissimilarities among assemblages were generated from a Euclidean distance matrix using latitude and longitude as dummy variables. (A) Dendrogram showing results of a cluster analysis of assemblages; linkages were based on weighted pair group averages; (B) distance map of the same assemblages produced by multidimensional scaling. NGF, northern gulf; DSDF, depauperate southern-derived fauna; SGF, southern gulf.

Figure 4.3. Common reef elements from the Gulf of California. (A) A *Fungia* spp. bed at Isla Monserrat; (B) a specimen of *Phyllangia consagensis* from Bahía Los Angeles; (C) a typical reef coral landscape in the Cabo Pulmo reef dominated by *Pocillopora* spp. Photo credits: (A) Andrés González-Peralta, (B, C) Israel Sánchez-Alcántara

26°N on the peninsular coast. Coral richness is much higher there than in the northern and eastern gulf. Although no actual framework exists, colonies of *Pocillopora* and *Pavona* can grow several meters tall. The corals are important to the ecosystem, because they provide substrate and protection for many other species. They also represent a key element at the base of the trophic web, because their zooxanthellae fix high amounts of carbon from photosynthesis and the corals, in turn, produce lipid-rich mucus for invertebrate and fish consumption (Muller-Parker and D´Elia 1997). For these reasons scleractinians offer a specific kind of habitat favoring the appearance of associated species, which would not otherwise occur

so abundantly in the gulf, including specialized corallivores such as the gastropod *Quoyula monodonta*, decapod crab *Trapezia* sp., and crown-of-thorns starfish *Acanthaster planci* (Reyes-Bonilla 2003). In addition, these communities are joined by an abundant and rich collection of free-living invertebrates that are generalists (Hendrickx et al. 2005). Especially conspicuous are echinoderms and mollusks, including species such as the brown urchin (*Tripneustes depressus*), black urchin (*Diadema mexicanum*), mother-of-pearl (*Pinctada mazatlanica*), and carnivore *Conus* and muricid snails. Correspondingly, the fish community is very complex (Thomson et al. 2000), including up to 25 species and over 300 individuals per 0.01-ha zones (Alvarez-Filip et al. 2006). Our personal observations indicate that in coral communities from protected areas where fishing has not been intense (such as Isla Montserrat), the teleost assemblage reaches a high trophic level (3.9 or more) due to the presence of carnivores such as groupers, snappers (*Epinephelus, Mycteroperca, Lutjanus*), grunts (*Haemulon* spp.), and pelagic species (jacks, barracuda, and others).

There has been much debate about the existence of actual coral reefs in the Gulf of California, because conditions are very limiting for coral growth and maintenance (Glynn 2001) and even the most extensive areas and those with the highest coral cover are poor in coral species and exhibit limited framework development when compared to the Caribbean or central and western Pacific reefs (Brusca and Thomson 1975). Because many other eastern Pacific reefs have environmental and ecological characteristics similar to those of the gulf (Cortés 2003), we propose that in the eastern Pacific, a "true reef" is a place where reef corals have a vital role in the food web by supplying resources to invertebrate and fishes as a result of photosynthesis performed by zooxanthellae, produce significantly higher relief than the surrounding rocky bottom, and provide a new type of habitat to be used by species with particular adaptations, which consequently become exclusive residents of these areas.

True coral reefs are quite limited in extent in the Gulf of California. Reyes-Bonilla (2003) noted two places: Bahía San Gabriel at Isla Espíritu Santo east of La Paz (24°N) and the well-known Cabo Pulmo reef (23.5°N). Both areas register the highest coral cover in the gulf (fig. 4.3B). Skeletal frameworks can reach over 2 m in height, and are predominantly built by *Pocillopora* sp. Considering that the growth rate of pocilloporids is about 3 cm/year in the gulf (Reyes-Bonilla and Calderón-Aguilera 1999), we estimate these colonies represent at least 70 years of continuous development.

Of the commonly cited reefs, only scant information is available on the associated communities of the Bahía San Gabriel reef (Squires 1959; Pérez-España et al. 1996). The Cabo Pulmo reef is described in much more detail, and there are dozens of studies on related topics, including marine species composition, geology, and ecological interactions (Brusca and Thomson 1975; Reyes-Bonilla and Calderón-Aguilera 1999; Riegl et al. 2007). In general, both of these reefs present a relatively diverse coral fauna with more than 10 species each (Reyes-Bonilla 2001). The interstices of the ramose colonies harbor a remarkable assemblage of macroinvertebrates (especially decapod crustaceans, mollusks, and echinoderms; Baynes 1999) and fishes (such as the coral hawk [*Cirrithichthys oxycephalus*], gobies, and moray eels; Villarreal-Cavazos et al. 2000). In addition, massive corals host a wide array of borers, including sponges (*Cliona* sp.), sipunculans, polychaetes, and bivalves (*Lithophaga* spp.), as well as several species of blennies (the dominant being *Acanthemblemaria crockeri*).

Coral stands offer protection to many other larger species, which usually live in the interfacing coral-sand or coral-rocky bottom. The most common such residents are urchins (*Tripneustes, Diadema, Eucidaris*), asteroids (*Pharia pyramidata*), mollusks (*Conus, Pinctada*), and fishes. Of the fishes, the herbivores *Stegastes* sp. and *Abudefduf troschelli*, the omnivore *Thalassoma lucasanum*, and the planktivore *Chromis atrilobata* represent 60 to 70 percent of the total abundance (Thomson et al. 2000; Alvarez-Filip et al. 2006). Carnivores are much less common and the key fish species are serranids (*Epinephelus, Mycteroperca,* and *Paranthias colonus*) and lutjanids (*Lutjanus viridis* and *L. argentiventris*).

Coral Community Structure

Data on abundance, richness, and other descriptors of coral-community structure in the Gulf of California are scarce. Reyes-Bonilla and Calderón-Aguilera (1999) provided the initial information about Cabo Pulmo reef. Here, we present new and updated information obtained between 2004 and 2007 from seven regions around the gulf: Bahía Los Angeles (28°N), Islas Tortuga and San Marcos (27°N), Bahía Concepción (26°N), Loreto (25°N), La Paz (24°N), Cabo Pulmo (23.5°N), and Los Cabos (23.1°N). Data were obtained in transects 25-m long, at depths from 6 to 12 m, and using the intercept point method. Divers took note of coral occurrences every 20 cm (making 100 total point counts). We conducted 24 transects at each site, with the exceptions of Tortuga and San Marcos (20 transects) and Bahía Concepción (18 transects). Information on amount of cover and richness was obtained directly from these transects and subsequently used to calculate diversity (Shannon-Wiener index *H'*, with base 10). According to a priori tests, all indices were normal or homoscedastic and, thus, we applied one-way parametric analysis of variance to test for differences among sites (factors). Note that field surveys were also conducted at Mazatlán, Sinaloa (20°N), and four areas of the Sonoran coast, from 27° to 31°N

(Guaymas, southern Isla Tiburón, Puerto Libertad, and Puerto Peñasco). However, as there was only one coral species present at these sites (*P. panamensis*) and cover did not exceed 1 percent, these data were not analyzed for this report.

Our surveys found that coral cover is less than 2 percent in the northern gulf, but increases rapidly in the southern gulf to a maximum level at Cabo Pulmo reef (fig. 4.4A). Conditions at Cabo Pulmo are adequate for coral growth due to abundant shallow hard substrate and because the water is warm most of the year (Reyes-Bonilla 2001, 2003). Cover did not increase monotonically with latitude, but dropped a bit at Los Cabos, perhaps because that area is characterized by a steep bottom and lack of available substrate related to a very narrow shelf. The ANOVA ($F_{5,152}$ = 12.4, P < 0.01) and an a posteriori Tukey test applied to coral-cover data indicate three groups: areas with high abundance (Cabo Pulmo, La Paz), medium (Los Cabos, Loreto), and low (the northernmost sites). Species richness behaved in a similar manner (fig. 4.4B), where Cabo Pulmo had the highest number of species per transect, followed by Los Cabos, La Paz, and Loreto, and finally by the areas at 27° and 28°N ($F_{5,152}$ = 4.6, P < 0.05).

On the other hand, diversity showed a slightly different picture (fig. 4.4C). Diversity was highest in the La Paz and Loreto areas. Despite the statistical test that suggests the four southernmost areas do not differ from one another, all are significantly more diverse than the assemblages from the Tortuga–San Marcos and Bahía Los Angeles areas ($F_{5,152}$ = 8.3, P < 0.05). Our results can be explained considering that Cabo Pulmo and Cabo San Lucas have a dominant species that unbalances the relative abundance of the taxa and, hence, decreases diversity. That species is *P. verrucosa*, which accounts for over 50 percent of the coral abundance in many transects in those two areas. At La Paz and Loreto, there is no single dominant coral, and instead relative abundances vary among sampling sites. Sometimes *P. verrucosa* appears in almost monospecific stands, but in many places south of Loreto and at Isla Espíritu Santo, *Pocillopora damicornis* is more common.

In summary, it is clear that while reef corals can be found practically in all rocky areas of the Gulf of California, their assemblages are quite simple everywhere except along the southernmost Baja peninsula. Patterns shown by ecological indices reflect the geographic regionalization cited earlier in this chapter. Latitudinal composition changes recognized in the

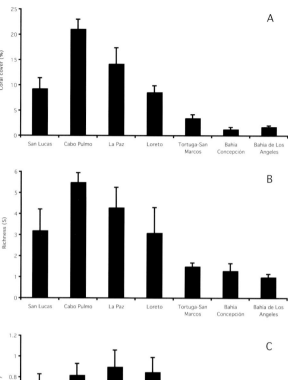

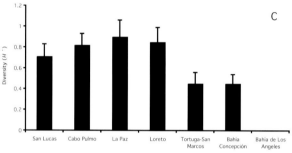

Figure 4.4. Community structure of selected coral reefs and assemblages in the Gulf of California: (A) coral cover; (B) species richness; (C) species diversity.

coral fauna appear to be mirrored in the structure and function of the assemblages.

Coral Reproduction

Reproduction and recruitment are among the most critical processes upon which the persistence of a coral reef depends, along with their pivotal influence on the structure and dynamics of coral populations and communities (Richmond 1997). Studies on eastern Pacific coral reproduction are scarce (Birkeland 1977; Richmond 1985, 1987; Glynn et al. 1991, 1994, 1996) and were only recently initiated in Mexico (Vizcaíno-Ochoa 2003; Mora-Pérez 2005; Rodríguez-Troncoso 2006; Chávez-Romo and Reyes-Bonilla 2007).

As elsewhere in the eastern Pacific (Glynn et al. 1994; Vizcaíno-Ochoa 2003; Rodríguez-Troncoso 2006),

Table 4.1. Carbonate deposition (kg CaCO$_3$.m^{-2}.yr^{-1}) at selected coral reefs and reef assemblages in the Gulf of California.

Locality	Coral cover (2001)	kg CaCO$_3$·m^2·y^{-1}	
		Min	Max
Isla San José	49.3	6.3	12.4
Bahía San Gabriel	35.8	5.8	13.5
Bahía Pichilingue	22.9	3.5	8.2
Isla Cerralvo	2.0	0.2	0.6
Punta Perico	8.6	1.3	3.0
Cabo Pulmo	53.6	8.9	20.7
Bahía Chileno	40.9	6.4	15.2

P. panamensis is a year-round brooder in the Gulf of California. Its gametogenic cycles and planulation show no correlation with sea-surface temperature or lunar activity, although fecundity and planula larval release is closely tied with sea-surface temperature (Mora-Pérez 2005). Active year-round larval release of *P. panamensis* populations in Gulf of California waters (Mora-Pérez 2005) probably maintain high recruitment and rapid population buildups similar to those in Central America (Glynn et al. 1994) and Oaxaca (López-Pérez et al. 2007). This level is maintained even against a high postrecruitment mortality rate (8–14 percent) and nonexistent recruitment via fragmentation (Reyes-Bonilla and Calderón-Aguilera 1994). Contrary to studies from Bahía de Banderas (Vizcaíno-Ochoa 2003) and Oaxaca (Rodríguez-Troncoso 2006) on mainland Mexico, work in the Gulf of California represents the first positive sighting of reproductively active *P. damicornis* populations in the Mexican Pacific region (Chávez-Romo and Reyes-Bonilla 2007). In the gulf, *P. damicornis* is a seasonal (July–November) broadcast-spawner, whose gametogenic cycles correlate with the sea-surface temperature. In particular, a high proportion of mature gametes result when the temperature exceeds 28°C (Chávez-Romo and Reyes-Bonilla 2007).

Coral Symbioses

The Gulf of California is the only area along the Pacific coast of Mexico where symbioses-related studies other than bleaching have been conducted. Reef corals are known to occupy the entire photic zone but display vertical zonation patterns within the light-intensity gradient (Wellington 1982). In the Gulf of California, shallow environments (0–6 m) are usually occupied by *P. verrucosa*, whereas deeper areas (6–14 m) are inhabited by *P. gigantea* (Reyes-Bonilla and López-Pérez 1998). The distinctive zonation pattern was previously explained by a combination of physical and biological factors affecting the coral host, including selective predation by fishes (Wellington 1982). Nonetheless, in situ measurements, transplant experiments, and molecular analysis have demonstrated that *P. verrucosa* harbors a dinoflagellate symbiont clade D, whereas *P. gigantea* hosts a type from clade C, each adapted to a particular light regime (Iglesias-Prieto et al. 2003, 2004; LaJeunesse et al. 2007). These findings suggest that the differential use of light by specific symbiotic dinoflagellates constitutes an important axis for niche diversification and is sufficient cause to explain the vertical distribution patterns of these two coral species (Iglesias-Prieto et al. 2004).

Carbonate Accretion

The influence of bioeroders on eastern Pacific carbonate accretion was investigated for Central American reefs, particularly after the 1982–1983 El Niño demonstrated the importance of biotic factors on the survival and evolution of these coral reefs (Glynn 1988). Changes in carbonate accretion have recently gained attention due to potential impacts from global climate. Nonetheless, eastern Pacific studies, particularly those performed in the Gulf of California, are precluded because historical records date only from 1987 and are restricted to Cabo Pulmo (Calderón-Aguilera et al. 2007). We calculate coral accretion for seven gulf localities (table 4.1) following the method used by Chave et al. (1972) and Reyes-Bonilla and Calderón-Aguilera (1999). Carbonate deposition is highly heterogeneous in the gulf in response to species composition, coral coverage, and reef dimension. Coral standing stock is high at Cabo Pulmo, Bahía Chileno, and Isla San José but extremely low at Isla Cerralvo, where 0.2 kg $CaCO_3/m^2$/year is deposited. We lack data for reef extension at places other than Cabo Pulmo, therefore precluding carbonate deposition estimates per reef system.

Cabo Pulmo estimates suggest this 150-ha reef system contributes as much as 13,482 to 31,023 tons of $CaCo_3$/ha annually. Yet, the relative importance of Cabo Pulmo carbonate deposition has been constantly decreasing since 1987 (Calderón-Aguilera et al. 2007, fig. 1) in response to coral disturbance (Reyes-Bonilla et al. 2002; Iglesias-Prieto et al. 2003).

Natural Disturbances

Corals are exposed to various natural agents of perturbation in the Gulf of California, the most relevant being the effects of predation, hurricanes, and warming caused by the El Niño Southern Oscillation (ENSO).

Coral predation is ubiquitous in the gulf, especially in Baja California Sur. The key coral consumers are fishes (in particular the spotted pufferfish [*Arothron meleagris*] and parrotfishes of the genus *Scarus*), gastropods (*Jenneria pustulata* and *Quoyula monodonta*) and echinoderms (crown-of-thorns starfish [*A. planci*] and occasionally echinoids like the pencil and brown urchins *Eucidaris thouarsii* and *T. depressus*; Reyes-Bonilla and Calderon Aguilera 1999; Reyes-Bonilla 2003). These species often bite the tips of *Pocillopora* and the sides of the massive *Porites* and *Pavona*, but apparently have a marked preference for the emerald coral, *P. panamensis*. In a series of fifteen 10-×-1-m belt transects run at Punta Galeras (near La Paz) in 2007, where the percentage of colonies with evidence of predation was evaluated, we found 72 ± 8 percent of the colonies of this species showed damage, compared to 12 ± 3 percent of *Pavona* and just 7 ± 1 percent of *Pocillopora*. The predilection of corallivores to attack *Porites* may be due to the fact that the skeleton of this coral is relatively porous and the tissue is embedded more than 0.5 cm inside. Hence, the amount of energy provided per bite may be higher for consumers.

There is only one estimate of the joint effect of corallivore species on coral reefs at Cabo Pulmo.

Reyes-Bonilla and Calderón-Aguilera (1999) showed that during the 1990s, when coral cover was very high (over 30 percent), predators ate less than 15 percent of the coral standing stock. Resampling of the same species showed that their numbers have not increased appreciably, to date, compared to a decade ago (Reyes-Bonilla et al. 2005b; Alvarez-Filip et al. 2006; unpublished data of the authors). Nevertheless, the effect of predators may be greater than before, due to a decrease in cover by approximately one-third after the 1997 ENSO bleaching event (Reyes-Bonilla 2001).

Hurricane damage to coral communities is restricted to southern Baja California, where the most conspicuous assemblages are established. Hurricane season lasts from July to November, but strikes are infrequent. According to the NOAA National Hurricane Center (http://www.nhc.noaa.gov/pastall.shtml), each year an average of 0.80 ± 0.22 hurricanes or tropical storms move over the area where reef coral communities and reefs are located along the peninsular coast. Reyes-Bonilla (2003) indicated that these events are capable of breaking large amounts of ramose colonies (*Pocillopora* and *Psammocora*) and can fragment massive colonies (*Pavona* and *Porites*). The *Fungia* banks also have been impacted. The largest *Fungia* bank, located near Isla San José in northern Bahía La Paz (25°N), was severely affected in 2001 and then in 2006, when Hurricanes Juliette and John shifted large amounts of sand away from the mainland and buried the corals. It is interesting that the rest of the ecosystem appears to have been remarkably resilient to major storm events. For example, no statistical change was detected in sea urchin abundance at Punta Arenas and San Gabriel (near La Paz) after Hurricane Isis in 1998 (Reyes-Bonilla 2003) and in urchins and reeffish numbers at Cabo Pulmo in 2006 after two major hurricanes (unpublished data). Lirman et al. (2001) reported a similar situation for the Huatulco coral reef tract in the Mexican Pacific.

By far, the most important perturbations suffered by reef corals were caused by the increase in sea-surface temperature brought by the ENSO. This phenomenon caused at least two major episodes of coral bleaching. The first event lasted from July to September 1987, but was relatively minor as coral mortality did not exceed 10 percent of the standing stock and no other reef taxa suffered damage (Reyes-Bonilla 1993b). A decade later, however, the eastern Pacific was the scenario of the strongest ENSO of the twentieth century (Wang and Fiedler 2006). Loss of coloration of the colonies began in July and was observed through November. All coral genera were affected, although *Pocillopora* suffered the most as a shallow-water resident (where warming was more intense), and their zooxanthellae are not physiologically adapted to conditions of very high temperature and light intensity (Iglesias-Prieto et al. 2004). The bleaching caused a loss in coral cover that averaged 18 percent in the Gulf of California (fig. 4.5), but as bad as that was, the gulf region was the least impacted in western Mexico. Coral reefs at Bahía Banderas (20°N) and Oaxaca (16°N) suffered over 50 percent mortality (Reyes-Bonilla et al. 2002). Based on this relationship, Reyes-Bonilla (2001) and Riegl and Piller (2003) suggested that the permanent front at the entrance of the gulf works as a buffer against extreme rises in sea-surface temperature.

A few years after the 1997 ENSO, Reyes-Bonilla et al. (2002) suggested that recovery of the populations would take about a decade, considering the timing in succession after the 1987 bleaching. We were mistaken. So far, corals have not recovered in the gulf as indicated by changes in coral cover in three reefs or coral communities of the southwestern gulf (fig. 4.5). In general, we infer that recovery was hampered by the continuous turmoil caused by hurricanes and tropical storms, but other, local effects also are detected. For example, corals at Bahía Chileno (23°N, north of Cabo San Lucas) were not so affected by the ENSO (Reyes-Bonilla 2001; fig. 4.5), but cover still has been in a gradual decline. The most likely reason is a consequence of the combined effect of storms and human perturbations, because the bay is among the most visited areas for diving and snorkeling in the state. On the other hand, the reef at Cabo Pulmo is almost free of anthropogenic disturbances but the impact of the ENSO was enhanced by several hurricanes (Alvarez-Filip et al. 2006; fig. 4.5). Coral abundance after 1997 is significantly less than before the ENSO ($F_{2,58} = 6.7$, $P < 0.001$; fig. 4.5). Finally, Bahía San Gabriel has less live coral in 2007 than a decade before, but the difference has never been significant (fig. 4.5). This reef turns out to be much more resistant than other reefs. The reason may be that pocilloporid corals are the dominant species and asexual reproduction at that locality is especially successful because the bay is shallow, closed, and the broken branches can settle on sand and stop moving (a circumstance that improves survival rate).

Human disturbances to corals and coral reefs in the Gulf of California are intense (Brusca et al. 2005). Fortunately for corals, however, the problem

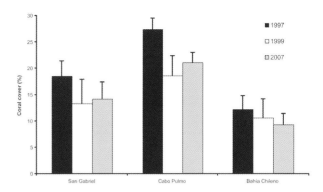

Figure 4.5. Change in coral cover in three coral reefs of the southwestern Gulf of California between 1997 (before the onset of the El Niño Southern Oscillation) and 2007.

is relatively minor in the southwestern corner of the region, where reefs are best developed and the human population has remained small. In fact, in 2005 the state of Baja California Sur had 512,000 inhabitants, according to the Instituto Nacional de Estadística, Geografía e Informática (http://www.inegi.gob.mx/est/), and was the least densely populated state in Mexico. Fishing activities during much of the last century did the most harm to coral faunas. Although no major damage was apparent by the 1980s, Sala et al. (2004) concluded that fishermen in the southern gulf had "fished down the food webs," because during the last 40 years they have taken fewer carnivorous fishes and more omnivores and herbivores as a result of depletion of the populations. Also, mean body sizes are noticeably smaller. Later, Sáenz-Arroyo et al. (2005) documented how the perception of the health of the rocky and coral ecosystems differs among fishermen of different ages, a clear example of so-called shifting baselines. Finally, personal observations by the authors indicate that sizes of typical commercial species (especially snappers and groupers) have increased remarkably at Cabo Pulmo reef after the establishment of the national park to protect the reef.

In general, the findings cited here indicate that extractive activities are having a measurable effect on the abundance and composition of reef fishes in traditionally fished areas. It is still uncertain, however, if those changes are affecting the function of the coral reefs or the ecosystem as a whole. Díaz-Uribe et al. (2007) developed a model for the food web in the La Paz area, where many coral communities are present. They showed that even when particular resources such as snappers, groupers, and sharks are overexploited, the ecosystem is not appreciably impacted. For example, all the artisanal and commercial fisheries together require less than 10 percent of the primary production to remain at the same level. In addition, a long-term analysis of the Cabo Pulmo fish community (1987–2003) revealed that, although community structure actually has been altered, many species were replaced by others of the same trophic level, family, or genus. This finding has been interpreted as evidence of ecological redundancy in the icthyofauna (Alvarez-Filip and Reyes-Bonilla 2006). Conflicting evidence on the effects of fishing on the ecosystem calls for increased research efforts in the future. This is an important issue, considering that any decision to stop or continue fishing at the current level will affect the economics and quality of life for many people in western Mexico.

The other major potential anthropogenic threat to coral reefs in the Gulf of California is tourism (Enríquez-Andrade et al. 2005), an activity so demanding of services that Baja California Sur grows by almost 3 percent yearly, the second-highest rate in Mexico (Lluch-Cota et al. 2006). Available reports indicate that so far the effects of visitors on coral reefs and communities seem to be minor, because most divers seek charismatic megafauna (hammerhead sharks, mantas) and choose to visit deeper sites away from the coast, while sport fishermen usually look for pelagic species (Bryant et al. 1998; Reyes-Bonilla 2003). In addition, the two most important coral areas in the gulf (Cabo Pulmo and Bahía San Gabriel) are protected areas (Arizpe-Covarrubias 2005; Anonymous 2007). Nonetheless, the speed of growth of tourist infrastructure and facilities (such as marinas and hotels) is staggering, and many investors are focused on the development of the Los Cabos–La Paz region, the so-called Los Cabos Touristic Corridor, precisely where the best coral communities and reefs are located. Taking this into account, the future of the coral-reef ecosystem depends on adequate management of the coastal zone and probably also on the relative success of low-impact tourism. This activity has been criticized as not meeting expectations of local managers and not improving the social condition of the general population. Consequently low-impact tourism may not turn out to be sustainable (Tershy et al. 1999; López-Espinosa de los Monteros 2002). Ecotourism has maintained a low level of reef use in some areas like Cabo Pulmo, while at the same time empowering the residents and boosting their economic capacity (Arizpe-Covarrubias 2005). There is still much to be done, and we urge that economic valuations of reef services be performed, as well as precise estimations of diving carrying capacity.

Azooxanthellate Corals

Deep-water, azooxanthellate or ahermatypic corals were once considered to be minor players compared to reef-builders, but a series of studies conducted during the last decade has demonstrated their importance as key elements in shelf, slope, and abyssal ecosystems (Roberts and Hirschfield 2004). For example, species such as *Desmophyllum dianthus* and *Lophelia pertusa* construct bioherms at depths beyond 100 m in temperate areas of the Northern and Southern Hemispheres. These systems are the setting for very active fisheries (Fossa et al. 2002).

The ahermatypes are ecologically and morphologically very different from reef species (fig. 4.3B). To begin with, they have no symbiont dinoflagellates, and they obtain food by catching particles and organisms through tentacular action. Hence, light has no influence on their distribution (Veron 1995). Another important difference is that most ahermatypes are solitary and not colonial (Cairns 1994). Finally, they have a much higher tolerance to cold water than reef species. Consequently many azooxanthellate species are cosmopolitan, circumpolar, or live at great depths, although their growth rate is usually very slow (Roberts and Hirschfield 2004).

There is almost no biological or ecological information regarding azooxanthellate corals in the Gulf of California, but good data are available on species richness and distributions. The regional fauna is composed of 22 species, two more than known for the shallow-water corals (20 species), which sums to a regional total of 42 stony corals (table 4.2). The gulf ahermatypes can live from the intertidal zone to a depth of over 600 m (Reyes-Bonilla et al. 2005a). They have the ability to reside in a variety of habitats and on different types of substrate, but most dwell on hard bottoms. Some species, such as *Dendrophyllia oldroydae*, can be so abundant in dredgings around the Midriff Islands that the local existence of deep-water reefs is highly likely (Reyes-Bonilla et al. 2008). In addition, other species are capable of recruitment on shells and rocks in sandy areas. Upon reaching maturity, those individuals actively detach themselves by dissolving their peduncles, as observed for *Endopachys grayi* (Cairns 1989). In some environments in the northern gulf, there exists a remarkable symbiosis

Table 4.2. Distribution of stony corals in the Gulf of California. Data from Reyes-Bonilla et al. (2005), Medina-Rosas (2006), Reyes-Bonilla et al. (in press), and field observations by the authors (2006–2007).

Species	BCS	BC	SON	SIN	Biogeographic affinity
Family Pocilloporidae					
Madracis pharensis (Heller, 1868)	X	X			Amphiamerican
Pocillopora capitata Verrill, 1866	X				East Pacific
P. damicornis (Linnaeus, 1758)	X				Indo Pacific
P. elegans Dana, 1846	X				Indo Pacific
P. eydouxi Milne Edwards and Haime, 1860	X				Indo Pacific
P. meandrina Dana, 1846	X				Indo Pacific
P. verrucosa (Ellis and Solander, 1786)	X			X	Indo Pacific
Family Poritidae					
Porites sverdrupi Durham, 1947	X	X		X	Endemic
P. panamensis Verrill, 1866	X	X	X	X	East Pacific
Family Siderastreidae					
Psammocora brighami Vaughan, 1907	X				Indo Pacific
P. haimeana Milne Edwards & Haime, 1851	X				Indo Pacific
P. stellata (Verrill, 1866)	X				Indo Pacific
P. superficialis Gardiner, 1898	X				Indo Pacific
Family Agariciidae					
Leptoseris papyracea (Dana, 1846)	X				Indo Pacific
Pavona clavus (Dana, 1846)	X				Indo Pacific
P. duerdeni Vaughan, 1907	X				Indo Pacific
P. gigantea Verrill, 1869	X				East Pacific
P. varians Verrill, 1864	X				Indo Pacific
Family Fungiidae					
Fungia curvata (Hoeksema, 1989)	X	X			Indo Pacific
F. distorta Michelin, 1842	X				Indo Pacific
F. vaughani Boschma, 1923	X				Indo Pacific
Family Rhizangiidae					
Astrangia californica Durham and Barnard, 1952		X	X		East Pacific
A. cortezi Durham and Barnard, 1952		X	X		Endemic
A. costata Verrill, 1866	X				East Pacific
A. dentata Verrill, 1866	X				East Pacific
A. haimei Verrill, 1866	X	X	X		East Pacific
Coenangia conferta Verrill, 1870	X	X	X		East Pacific
Oulangia bradleyi Verrill, 1866	X	X			East Pacific

Table 4.2. *continued*

Species	BCS	BC	SON	SIN	Biogeographic affinity
Family Caryophylliidae					
Caryophyllia diomedeae Marenzeller, 1904		X			Indo Pacific
Ceratotrochus franciscana Durham and Barnard, 1952	X		X		Endemic
Coenocyathus bowersi Vaughan, 1906	X	X	X		East Pacific
Desmophyllum dianthus (Esper, 1794)		X			Cosmopolitan
Heterocyathus aequicostatus Milne Edwards and Haime, 1848	X	X	X	X	Cosmopolitan
P. stearnsii Verrill, 1869	X	X	X		East Pacific
P. consagensis (Durham and Barnard, 1952)	X	X			East Pacific
P. dispersa Verrill, 1864	X				East Pacific
Family Turbinoliidae					
Sphenotrochus hancocki Durham and Barnard, 1952	X				Indo Pacific
Family Dendrophylliidae					
Balanophyllia cedrosensis Durham, 1947	X	X	X		East Pacific
Cladopsammia eguchii (Wells, 1982)	X				Indo Pacific
Dendrophyllia oldroydae Oldroyd, 1924	X	X	X		East Pacific
Endopachys grayi Milne Edwards and Haime, 1848	X	X			Indo Pacific
Tubastraea coccinea Lesson, 1829	X				Indo Pacific

between the coral *Heterocyathus aequicostatus* and a sipunculan of the genus *Aspidosiphon*. The coral has a round opening at the base of the colony in which the worm takes shelter. In return, the worm is in constant motion like a muscular foot, an activity that keeps the coral from being covered by sand (Reyes-Bonilla et al. 2008).

Reyes-Bonilla and Cruz-Piñón (2000) and Reyes-Bonilla et al. (2005a) studied azooxanthellate coral patterns in western Mexico from a biogeographic perspective. Both studies demonstrate how the Gulf of California fauna can be differentiated from that of the rest of the country, and how the gulf is subdivided into three regions: south (22° to 26°N), central (27° to 29°N), and north (30° to 31°N). It also is remarkable that in complete contrast to the reef fauna (composed mostly of Indo-Pacific species), there are only three immigrant azooxanthellate species in the region (*E. grayi*, *Tubastraea coccinea*, and *Cladopsammia eguchii*); all belong to the family Dendrophylliidae. The rest are amphi-American in origin (*Madracis pharensis*), cosmopolitan (*D. dianthus*), and gulf endemics

(*A. cortezi* from the northern gulf and *Ceratotrochus franciscana* from the La Paz region), but there also are 16 eastern Pacific endemics.

In addition to general patterns, Reyes-Bonilla et al. (2008) noted the presence of two species with a disjunctive distribution: *Paracyathus stearnsii* and *Balanophyllia cedrosensis*. These corals occur in the northern gulf and along the Pacific coast of Baja California, but not in between (Reyes-Bonilla et al. 2005). Remarkably, they already show some morphological differentiation between these sites (Cairns 1994). It is possible that the observed changes are the first evidence of incipient speciation caused by geographic isolation, a process known to occur in several fish species showing the same distributional pattern (Bernardi et al. 2003).

One might think that because most azooxanthellate corals live at a water depth of more than 30 m, there is no need for concern about their conservation. The situation is potentially serious, however, because deep-water coral assemblages are severely impacted by fishermen in the eastern and western United States and Canada, as well as New Zealand, Japan, and Norway (Roberts et al. 2006). The reason is that azooxanthellate corals are extracted as incidental by-catch. The problem is so severe in Japan and Norway that an estimated 30 to 50 percent of the deep-water reefs are damaged.

In the Gulf of California, Reyes-Bonilla et al. (2008) indicated that several species have been affected by human activities. Particular attention was called to *E. grayi* and *H. aequicostatus*, which live on sandy, flat-bottomed parts of the upper continental shelf where the shrimp fishery operates (Steller et al. 2003). A recent evaluation of the status of the deep-water corals from western Mexico (Reyes-Bonilla et al. 2008) found that *Astrangia costata, C. franciscana, C. eguchii, Dendrophyllia californica*, and *Sphenotrochus hancocki* must be regarded as under threat according to the risk evaluation method established by the National Ecology Institute (Anonymous 2001). Unfortunately, no marine area away from the coastal margin is protected in Mexico, and that situation puts azooxanthellate corals at risk.

5 Coral Diversification in the Gulf of California During the Late Miocene to Pleistocene

Ramón Andrés López-Pérez and Ann F. Budd

The origin and development of an eastern Pacific coral fauna is explained by competing theories commonly identified as the dispersal hypothesis and the vicariance hypothesis. Championed by Dana (1975), the dispersal hypothesis claims all eastern Pacific corals that were mainly Atlantic in origin became extinct and were replaced by an Indo-Pacific fauna during the Pliocene-Pleistocene glaciations well after the formation of the Isthmus of Panama. In contrast, Heck and McCoy (1978) asserted that the modern eastern Pacific fauna was derived from a previous Atlantic fauna through vicariance after the rise of the Isthmus of Panama. Vicariance is favored, they argued, by the rarity of successful long-distance dispersal events and possible similarities between modern eastern Pacific corals and Caribbean corals of Tertiary age. Mollusks, crabs, echinoderms, and bryozoans are good examples of shallow-water, reef-associated invertebrates currently living in the eastern Pacific that bear strong affinities to Caribbean counterparts (Budd 1989), thus suggesting evolution from the same stock. Conversely, the dispersal hypothesis is supported by the similarity of modern eastern Pacific coral taxa and by a presumed lack of related taxa in the Caribbean Neogene (Dana 1975). Today, it is more widely accepted that the coral taxa changed from Caribbean to Indo-Pacific affinities between the Pliocene and the Recent (Vaughan 1917; Durham 1966; Heck and McCoy 1978; Budd 1989; Colgan 1990; Reyes-Bonilla 1992).

Although the general result—the biogeographic change from Caribbean to Indo-Pacific—is strongly supported, the faunistic transition has not been carefully addressed. It is unclear whether the transition involved complete faunal collapse and restructuring as opposed to replacement of individual species within pre-existing communities. Failure to properly address the faunal turnover and the origin of the eastern Pacific coral reef fauna is related to the relatively poor development of eastern Pacific reef buildups before, during, and after the turnover event; the lack of abundant well-preserved specimens; our inability to identify species and interpret their evolutionary relationship; and the lack of age constraints on eastern Pacific outcrops with fossil corals.

The Gulf of California is especially suitable for a reconsideration of questions regarding the evolution of the eastern Pacific coral fauna. This is due to the many Late Miocene to Recent coral outcrops reported in the area (Vaughan 1917; Jordan and Hertlein 1926; Hanna and Hertlein 1927; Durham 1947, 1950; Hertlein 1957; Hertlein and Emerson 1959; Squires 1959; Simian and Johnson 1997) and to the relative absence of fossil corals in western Mexico and Central America (Palmer 1928; Hertlein 1972). In this chapter we examine Late Miocene to Pleistocene coral assemblages in the Gulf of California. In studying the taxonomic composition and relative abundance of corals from reef sequences in this region, three sets of interrelated questions may be addressed. First, does taxonomic diversity show change within the Gulf of California over the past 7 Ma? Second, have distinct episodes of origination and/or extinction occurred within the gulf over the past 7 Ma? If so, over what time interval did accelerated extinctions and originations occur, and were periods of accelerated extinction and origination synchronous? Finally, how have reef communities changed over the past 7 Ma? Has turnover been episodic or gradual? Were new species added before, during, or after the demise of pre-existing communities?

Geological Setting

Coral compositions were determined from assemblages collected from outcrops with fossil corals that show reef development from the Late Miocene to Recent located in the Gulf of California area (fig. 5.1). The gulf is a marginal sea between the Baja California peninsula and northwestern Mexico in the eastern

Pacific. It evolved through a complex geological history commonly summarized in three developmental phases (see chap. 1). The first involved active subduction on the Pacific coast of California from about 25 to 12.5 Ma. The second phase entailed crustal extension related to the opening of the proto-gulf (10–3.5 Ma). The third phase saw a reorganization by 3.5 Ma from crustal extension to transtensional tectonics responsible for the configuration of the Gulf of California today. Age determinations of the oldest coral-bearing units deposited in the Imperial Valley (McDougall et al. 1999) and Isla Tiburón (Gastil et al. 1999) suggest that coral settlement and development occurred during later development (i.e., the extension and transtensional tectonic phases).

Coral-bearing units are represented in the form of low-angle ramps (Punta Chivato area, San Nicolás, Isla Monserrat) or flat-lying terraces (Isla Coronados, Las Animas, Cabo Pulmo; fig. 5.1; table 5.1) of variable extent usually resting with an angular unconformity on the tilted volcanics of the Miocene Comondú Group in the Bahía Concepción area (Ledesma-Vázquez and Johnson 2001), Miocene El Cien Formation at Las Animas (DeDiego-Forbis et al. 2004), and Late Miocene–Early Pliocene Trinidad Formation at Los Algodones (Martínez-Gutiérrez and Sethi 1997). Most outcrops are small and reminiscent of more extensive deposits from exposed, high-energy

environments; however, other units deposited within protected embayments also are present (South Punta Chivato, Cañada Coronados, Puerto Balandra, Las Animas; fig. 5.1; table 5.1). Marine-terrace deposits are highly common and widespread from Santa Rosalía to Cabo Pulmo (Ortlieb 1991). Those recording reef development, however, are scarce. Except at Isla Coronados and La Ventana, where multiple coral terraces occur, they represent single spatio-temporal growth episodes. Indo-Pacific and Caribbean reef systems have a wide bathymetric range (about 0–50 m) and, therefore, most reef-building episodes preserve several reef environments (Pandolfi 1996). In contrast, Gulf of California coral communities developed in shallow waters (< 15 m) and coral-bearing units generally represent single reef environments.

There has been a relatively large improvement in the age resolution of several Gulf of California outcrops (table 5.1), but dates are needed for recently sampled places where multiple growth episodes are recorded (Isla Coronados and La Ventana). At Isla Coronados and La Ventana, the stratigraphy and absolute age dates for some terraces (Sirkin et al. 1990; Johnson et al. 2007) allow us to establish the relative relationships for those areas. Precise age relationships among different places, however, are poorly known, and more effort is needed in this respect.

Studying Coral-Bearing Units
Sampling

Data were collected during four field expeditions to the Baja California peninsula in June–August 2002, January 2003, June–July 2003, and January 2005. For each coral-bearing unit under study, 1-m^2 (1 × 1 m) quadrats were systematically sampled. All quadrats were randomly placed in each coral-bearing stratigraphic horizon, and all coral species within the quadrat were recorded. Specimen collection followed two approaches at each locality: (1) within each quadrat, individual coral specimens were extracted from the face of the outcrop so that the collections would be qualitatively representative of the species composition and abundance; and (2) unrecorded species were directly collected during a thorough search of the locality. Due to variations in size from 1-m^2 quadrats (BC 22 at Isla San José) to hundreds of meters (BC 38 at Isla Coronados), in preservation (BC 34 at El Bajo), and in community complexity of the fossil-bearing units (i.e., mono- or multispecific communities), equivalent volumes of material were not collected

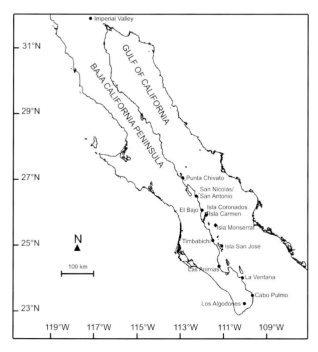

Figure 5.1. General location of faunas included in the analysis.

Table 5.1. List of Gulf of California zooxantellate coral collections sites arranged in stratigraphic order from oldest to youngest. Ma = million years.

Field number	Site name	Number of specimens	Number of species	Formation	Age (Ma)
USGS 3923	Alverson Canyon	44	4	Imperial	6–6.5 Ma[7]
USGS 7616	Barret Canyon	45	5	Imperial	6–6.5 Ma[7]
UCLA 631	Coyote Mountains	95	6	Imperial	6–6.5 Ma[7]
BC 3	Punta Chivato	88	3	San Marcos	E. Pliocene
BC 35	Ensenada El Muerto	51	7	San Marcos	E. Pliocene
BC 15	Puerto de La Lancha	51	5	San Marcos	E. Pliocene
BC 9	Los Algodones	31	1	El Refugio	E. Pliocene
BC 22-23	Isla San José	3	3	—	>3 Ma[9]
BC 4	San Nicolás	9	1	San Nicolas	<3.3 Ma[6]
BC 24-27	Isla Monserrat	109	3	Carmen	3.1–1.8 Ma[8]
BC 2	Las Barracas	13	2	Marquer	3.1–1.8 Ma[8]
BC 19	Bahía Marquer	14	1	Marquer	3.1–1.8 Ma[8]
BC 32	La Ventana-1	6	2	—	—
BC 31	La Ventana-2	3	2	—	—
BC 29-30	La Ventana-3	29	2	—	—
BC 11, 28	La Ventana-4	79	5	—	>0.3[5]
BC 44	Isla Coronados-1*	0	1	—	<0.16[10]
BC 43	Isla Coronados-2*	0	1	—	—
BC 42	Isla Coronados-3*	0	2	—	—
BC 41	Isla Coronados-4	34	4	—	—
BC 40	Isla Coronados-5*	0	2	—	—
BC 39	Isla Coronados-6	9	2	—	—
BC 14	Arroyo Blanco	1	1	—	—
BC 8	Cabo Pulmo	11	5	—	0.14–0.12[3]
BC 38	Cañada Coronados	18	3	—	0.121 ± 6[4]
BC 17-18	Puerto Balandra	22	1	—	—
BC 20-21	Timbabichi	57	5	—	—
BC 10	La Ventana-5	19	4	—	—
BC 12-13	Punta Baja	46	4	—	—
BC 16	Bahía Otto	6	2	—	—
BC 6-7, 33	Las Animas	37	4	—	0.13–0.128[1]
BC 5	San Antonio	8	1	—	0.13–0.12[2]
BC 1	South Punta Chivato	18	1	—	—
BC 34	El Bajo	17	1	—	0.12[11]
BC 37	Isla Coronados-8	6	2	—	—
BC 36	Isla Coronados-9	9	2	—	—

at each site. Therefore, data were lumped into more uniform and meaningful sampling units, hereafter referred to as "assemblages." Instead of coral coverage commonly used in community analysis, it was necessary to use the number of colonies per assemblage to address relative abundance.

Data Analysis

Species were identified by taking measurements directly on specimens and analyzing the data following a rigorous morphometric protocol designed to detect distinct morphologic entities through geologic time (Budd and Coates 1992; López-Pérez et al. 2003). In general, specimens were identified and distinguished following a suite of geometric morphometric and traditional morphometric analyses, comparing fossil and modern specimens previously obtained from the Gulf of California (Durham 1950; Durham and Barnard 1952; Squires 1959), eastern Pacific (Reyes-Bonilla et al. 2005a), and the Caribbean (Budd 1991; Budd et al. 1994; Budd and Johnson 1999a). Some identifications have been left in open nomenclature pending the formal taxonomic publication of the systematic work.

Because the age resolution of samples is not uniform (fig. 5.2), conservative estimates of species richness and taxonomic turnover were calculated following the methods used by Budd et al. (1996) and Johnson (2001). Each sample was assigned a "long" stratigraphic range extending from the lower age boundary of the sample in which the species first occurs to the upper age limit of the sample in which the species last occurs. Counts of range-through species richness and the numbers of first and last occurrences of species were then calculated from the stratigraphic ranges for a set of 0.5- and 1-Ma subintervals extending from the Late Miocene to the Recent. Both

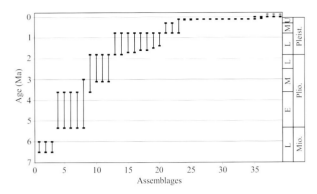

Figure 5.2. Stratigraphic ranges of 40 Late Miocene to Recent assemblages included in the analysis. Asterisks represent Recent assemblages. Miocene and Pliocene: L, Late; M, Middle; E, Early; Pleistocene: U, Upper; M, Middle; L, Lower.

conservative and nonconservative taxonomic turnover rates were estimated. Conservative rates were estimated by extending first and last species occurrences over several time intervals. For instance, a last occurrence at a locality with a resolution of two time intervals is treated as one-half extinction in each of the two intervals. This acts to smooth peaks that might otherwise result from using midpoints of sample age ranges to associate samples with a single subinterval in the absence of evidence that the true age of the sample was restricted to that subinterval (Johnson 2001). Speciation and extinction rates were estimated by dividing the number of first or last occurrences for each time interval by the total number of species living during that interval, and then dividing the proportion by the interval duration. Nonconservative taxonomic turnover rates were estimated, as above, except that first and last species occurrences were assigned to the first and last time interval in which they occurred.

Presence/absence values and the relative abundances of coral taxa were used to analyze community dynamics in the gulf. A relative abundance matrix was created by counting the number of specimens per species collected within each assemblage and assigning codes for rare (one specimen), common (two or three specimens), abundant (three to nine specimens), and super-abundant (more than nine). Presence/absence values and relative abundances of coral taxa among assemblages were compared using the Bray-Curtis dissimilarity index. This index was adopted because it is one of the most robust for analysis of relative abundance data (Clark and Warwick 2001; McCune and Grace 2002). The Bray-Curtis similarity matrix was

Table 5.1. NOTES

Total no. collections = 47; total no. specimens = 988 ; median no. specimens = 20.97 ; min-max no. specimens per collection = 1-103; total no. species = 23; median no. species per collection = 2.36; min-max no. species per collection = 1–7.

1 = DeDiego-Forbis et al. (2004); 2 = Johnson and Ledesma-Vásquez (1999); 3 = Muhs et al. (1994); 4 = Johnson et al. (2007); 5 = Sirkin et al. (1990); 6 = Ledesma-Vázquez (2002); 7 = Eberly and Stanley (1978), McDougall et al. (1999); 8 = Dorsey et al. (2001); 9 = Umhoefer et al. (2007); 10 = Bigioggero et al. (1988); 11 = Mayer and Vincent (1999); * not collected.

analyzed through a suite of agglomerative unweighted pair-group average cluster analysis (UPGMA) and multi-dimensional scaling (MDS). The hierarchical agglomerative cluster was used because of its ability to provide a visual summary for the patterns among assemblages and its demonstrated utility in ecological studies. Analysis by MDS was selected over other commonly used ordination techniques (e.g., principal component analysis and detrended correspondence analysis), because in many comparative ordination studies this technique is considered to be one of the more powerful ordination analyses available (Clark and Warwick 2001; McCune and Grace 2002). Both ordination and cluster analyses were performed because they provide a graphical display linking assemblages with mutually high levels of similarity that groups assemblages into discrete clusters. Due to the degree of arbitrariness in classification techniques, UPGMA and MDS were followed by analysis of similarity (ANOSIM), a simple nonparametric permutation procedure applied to the similarity matrix (Clark and Warwick 2001). The ANOSIM technique was applied to search for meaningful differences among group assemblages. The MDS and ANOSIM analyses were followed by a similarity percentage analysis (SIMPER; Clark and Warwick 2001) to determine the taxa responsible for the similarity within assemblages (Edinger et al. 2001).

Sampling and Taxonomic Descriptions

The number of specimens collected per locality ranges from 1 to 103 (median = 20.97), and the number of species collected per locality ranges from 1 to 7 (median = 2.36; table 5.1). Distribution frequencies for number of both specimens and species per locality are platykurtic and skewed to the right. This resulted from the high number of localities containing low numbers of species and specimens (fig. 5.3A, B). The relationship between the number of specimens per locality and number of species per locality show a stepped plateau shape with multiple sharp changes near a locality having 20 specimens and four species and a locality having 45 specimens and five species. This outcome suggests the presence of groups of localities with approximately the same number of species (fig. 5.3C).

Cumulative species curves were constructed by adding localities in temporal order beginning with the stratigraphically oldest and continuing to the

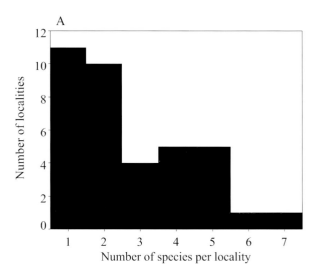

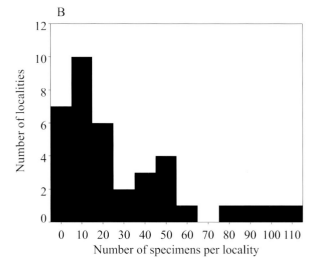

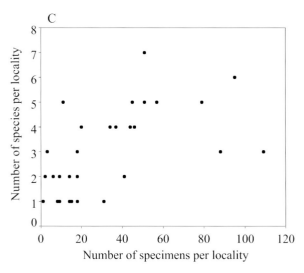

Figure 5.3. Histograms and scatterplot showing the numbers of species and specimens collected per locality.

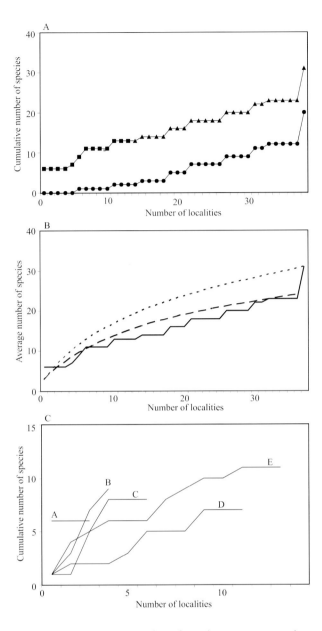

Figure 5.4. Cumulative number of species curves assessing sample adequacy. (A) Curve constructed by adding localities in stratigraphic order. Each point represents an assemblage. Squares, pre-turnover taxa; triangles, post-turnover taxa; circles, post-turnover taxa alone. (B) Curve constructed by randomly resampling 999 assemblages. Continuous, pre-turnover and post-turnover taxa; largely dashed line, random curve, Recent excluded; finely dashed line, random curve, all assemblages included. (C) Cumulative curves per age. A, Late Miocene; B, Early Pliocene; C, Middle to Late Pliocene; D, Lower to Middle Pleistocene; E, Late Pleistocene.

youngest for three kinds of data: (1) species without living representatives, "pre-turnover species"; (2) species with living representatives, "post-turnover species"; and (3) pooled pre-turnover + post-turnover species (fig. 5.4A). In general, the curves display a stepped shape with the steps strongly corresponding with the age of the sampled formations. For instance, the first plateau corresponds to the Late Miocene Imperial Formation of south-central California (6.5–6 Ma), the second to the Early Pliocene San Marcos and El Refugio Formations (5.3–3.6 Ma), whereas the last plateau corresponds to the Upper Pleistocene of Baja California (about 125 Ka). This result seems to suggest that the species living within each formation/age were adequately sampled. Nonetheless, figure 5.4B shows that the sampling did not reach saturation in the Baja California peninsula area, because neither the randomly resampled curve including the Recent localities nor the curve without the Recent localities levels off. This is best depicted in figure 5.4C, where except for the relatively well-developed plateau for the Middle to Late Pliocene Carmen/Marquer Formation (3.6–1.8 Ma), the Early to Middle Pleistocene (1.8 Ma–126 Ka) and Late Pleistocene (about 126 Ka) curves do not level off. Additional collecting, especially within Upper Miocene and Lower Pliocene strata, is likely to result in the discovery of new taxa, especially after the inspection of new exposures from a wide range of environmental settings.

Comparison between pre-turnover and post-turnover curves (fig. 5.4A) suggests that from the Late Miocene until the final addition of *Favia* sp. A in the Middle to Late Pliocene of Carmen/Marquer Formation at Isla Monserrat (3.6–1.8 Ma), pre-turnover taxa were continuously added and dominated the Gulf of California fauna. Except for the common and widespread *Porites panamensis* (fig. 5.5, no. 1) and *Pocillopora capitata* that arose in the Lower Pliocene San Marcos Formation of Isla Carmen and the Middle to Upper Pliocene Carmen/Marquer Formation of Isla Monserrat, no other living species coexisted with pre-turnover taxa in the Gulf of California. Living species increased in number after the demise of pre-turnover taxa. The first occurrence of an Indo-Pacific species is marked by *Pavona clavus* in the Lower to Middle Pleistocene of La Ventana near the mouth of the gulf. Here, well-developed though rare *P. clavus* colonies are interspersed among abundant colonies of *P. panamensis*. These species do not form a reef structure and are not bound together to a firm substrate. Instead they are loosely cemented and surrounded by bioclastic sandstone and Quaternary siliciclastics. Finally, after the addition of *P. clavus* to the gulf fauna, a steady but sporadic arrival of Indo-Pacific species occurred.

Of the 23 identified species, 12 are living and 11 are extinct (table 5.2). The 12 living species represent approximately 75 percent of the 16 hermatypic species in the Gulf of California and about 27 percent of 44 eastern Pacific species (Reyes-Bonilla 2002). Among the living species are several of the major reef-builder species in the Gulf of California and Mexican Pacific (Reyes-Bonilla and López-Pérez 1998; Reyes-Bonilla et al. 2002; López-Pérez and Hernández-Ballesteros 2004) and eastern Pacific communities (Glynn and Ault 2000). They include *Pocillopora damicornis* (fig. 5.5, no. 2) *P. capitata, P. panamensis, Porites lobata* (fig. 5.5, no. 5), *Pavona gigantea, Gardineroseris planulata* (fig. 5.5, no. 4), the more restricted but common *Psammocora stellata, P. clavus*, and the Gulf of California endemic *Porites sverdrupi*. Among the extinct corals are the common *Solenastrea fairbanksi, Porites carrizensis, Dichocoenia merriami, Diploria bowersi*, and *Siderastrea mendenhalli*, as previously reported from the Imperial Formation and scattered places in the Gulf of California (Vaughan 1917; Jordan and Hertlein 1926; Simian and Johnson 1997; Gastil et al. 1999; López-Pérez 2005). Also included are well-preserved specimens of *Dichocoenia eminens* (fig. 5.5, no. 11) from the Early Pliocene–Early Pleistocene on the northern Caribbean and *Diploria sarasotana* (fig. 5.5, no. 12) from the Early Pliocene–Late Pliocene Tamiami Formation of Sarasota, Florida, also in the northern Caribbean. Finally, four of the species identified in the collections as *Placosmilia*? sp. A (fig. 5.5, no. 8), *Siderastrea* sp. A (fig. 5.5, no. 7), and two species of *Favia* (fig. 5.5, nos. 3, 10) are undescribed and will be formally named and described elsewhere.

Of the 23 identified species, eight are temporally restricted to one assemblage in the Gulf of California. Of these, four are restricted to the Pliocene, one to Middle Pleistocene, and the rest to the Late Pleistocene. Nonetheless, *G. planulata* and *P. lobata* also occur in the Recent of the Mexican Pacific, whereas *D. eminens* and *D. sarasotana* occur in the Pliocene of the Caribbean (http://nmita.geology.uiowa.edu). Of the pre-turnover species, *S. mendenhalli* is restricted to the Late Miocene Imperial Formation, whereas those previously identified as *P. carrizensis, D. bowersi, S. fairbanksi*, and *D. merriami* are much more spatio-temporally distributed than previously thought (Vaughan 1917; Jordan and Hertlein 1926; Hertlein

and Emerson 1959; Simian and Johnson 1997; Gastil et al. 1999). *Pocillopora capitata*, which is first recorded in the Middle to Upper Pliocene undifferentiated Carmen/Marquer Formation at Isla Carmen, was then spatially restricted to several Isla Coronados terraces but flourished all over the Gulf of California by the Late Pleistocene. Finally, since its first appearance in the Lower Pliocene San Marcos Formation of Isla Carmen, *P. panamensis* achieved a wide spatio-temporal distribution in the gulf and became the major reef builder during most of the Pleistocene.

Faunal Turnover

Estimates of species richness increased slightly during the Early Pliocene and remained relatively constant until the Middle Pliocene, when a drop in species occurred. During the drop, richness reached its

Figure 5.5. (1) *Pavona clavus*. Figured specimen. SUI 100866. Late Pleistocene, locality BC 11, La Ventana, Baja California Sur, Mexico. (2) *Pocillopora damicornis*. Figured specimen. SUI 100626. Late Pleistocene, locality BC 8, Cabo Pulmo, Baja California Sur, Mexico. (3) *Favia* sp. A, new species. Holotype. SUI 100686. Middle Pliocene, Carmen Formation, locality BC 27, Isla Monserrat, Gulf of California, Mexico. Transverse section. Corallite detail. (4) *Gardineroseris planulata*. Figured specimen. SUI 100660. Middle Pleistocene, locality BC 11, La Ventana, Baja California Sur, Mexico. (5) *Porites lobata*. Figured specimen. NHMLAC 11739. Late Pleistocene, locality BC 8, Cabo Pulmo, Baja California Sur, Mexico. (6) *Pocillopora meandrina*. Figured specimen. NHMLAC 11740. Late Pleistocene, locality BC 8, Cabo Pulmo, Baja California Sur, Mexico. (7) *Siderastrea* sp. A, new species, San Marcos Formation. SUI 100673. Early Pliocene, San Marcos Formation, locality BC 15, Puerto de la Lancha, Isla Carmen, Gulf of California, Mexico. (8) *Placosmilia*? sp. A, new species. Holotype. SUI 100680. Middle Pliocene, San Nicolás Formation, locality BC 4, San Nicolás, Baja California Sur, Mexico. (9) *Porites panamensis*. Figured specimen. SUI 100677. Late Pleistocene, locality BC 33, Las Animas, Baja California Sur, Mexico. (10) *Favia* sp. B, new species. Holotype. SUI 100688. Early Pliocene, San Marcos Formation, locality BC 15, Puerto de la Lancha, Isla Carmen, Gulf of California, Mexico. (11) *Dichocoenia eminens*. Figured specimen. SUI 100612. Late Miocene, Imperial Formation, locality UCLA 631, Carrizo Creek, northeast Coyote Mountains, California, United States. (12) *Diploria sarasotana*. Holotype. UF 8279. Early Pliocene–Late Pliocene, Tamiami Formation; Sarasota, Florida. All scale bars represent 1 cm. Photo credit: Ramón Andrés López-Pérez

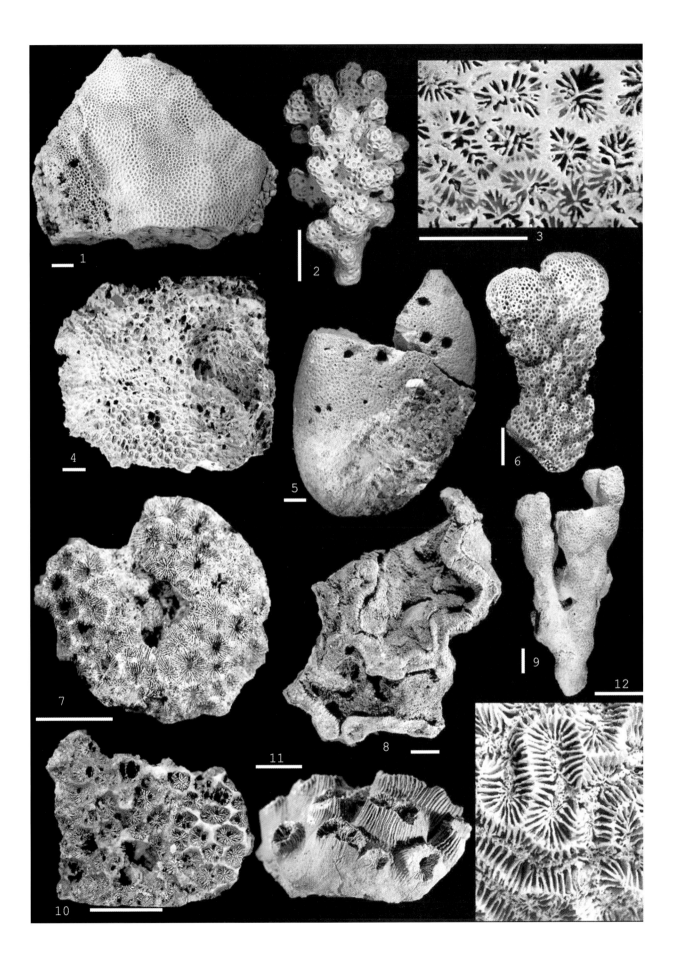

minimum during the Late Pliocene to Early Pleistocene. After the drop, species richness increased to reach its maximum during the Late Pleistocene (fig. 5.6A). This pattern is produced by the high origination rate concentrated during the Late Miocene–Early Pliocene and Late Pleistocene ages (fig. 5.6B). Extinction was relatively constant during the Pliocene, though a slight increase occurred at 3–1 Ma (fig. 5.6C), which was responsible for the drop in species richness during the same time interval. Conservative and nonconservative origination and extinction estimates yield approximately similar results coinciding with high and low estimates for both approaches, although nonconservative extinction estimates produce two distinct extinction pulses at about 3 and 1.5 Ma, which mimic more accurately the pattern of species loss.

Visual comparison of species richness and sampling intensity (i.e., number of assemblages) per time interval suggest a relationship between both variables. A Spearman rank correlation test reveals that both schemes of sampling (i.e., 0.5 and 1 Ma) fully explain species richness ($n = 14$, $R = 0.771$, $P = 0.016$), but except for estimates of conservative and nonconservative origination rates per 0.5 Ma, it does not explain the high species origination observed at 6.5–5 Ma nor the high species extinction observed at 3 and 1.5 Ma. Sampling intensity is especially tied to those values occurring during the last 2 Ma (fig. 5.6A).

Pattern of Faunal Change

Assemblage analyses of presence/absence and relative abundance data yield similar results with minor differences in the internal composition of the groups. Therefore, assemblage changes are explained on the basis of relative abundance data analysis alone. Cluster analysis shows the existence of two major groups corresponding mainly to Upper Miocene to Upper Pliocene assemblages (group A) and Pleistocene to Recent assemblages (group B; fig. 5.7A). Furthermore, a clear subdivision is detected within major groups. With respect to group A, the main division occurs between a mix of Imperial Formation and Las Barracas assemblages (A1) and a mix of the San Marcos Formation and El Peyote assemblages (A2); whereas Puerto de la Lancha (BC 15), Isla San José (A3), and San Nicolás (A4) represent unique assemblages of uncertain affinity. Within group B, the larger group of clustered assemblages mainly ranges from the Lower to Upper Pleistocene (B1), group B3 corresponds

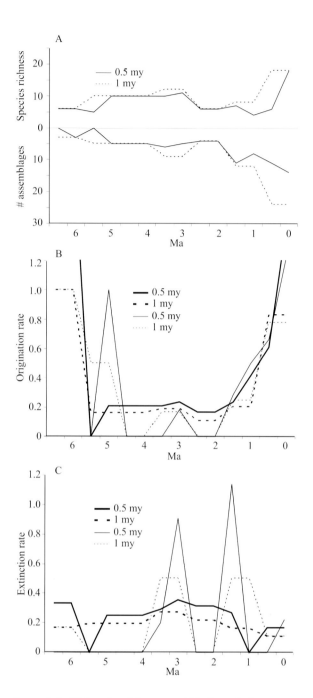

Figure 5.6. (A) Species richness and (B, C) evolutionary rates of Gulf of California reef corals from the Late Miocene to Recent. Each plot shows patterns calculated with interval durations of 0.5 and 1 Ma. Thick lines, conservative estimates; thin lines, nonconservative estimates.

with Upper Pleistocene assemblages restricted to the southern part of the Baja California peninsula area, whereas group B4 represents Recent Baja California assemblages (BCS C–S). Finally, the Upper Pleistocene Punta Baja (BC 12, 13), Middle Pleistocene La Ventana (BC 11, 28), and Middle to Late Pliocene Isla

Carmen (BC 24–27) assemblages represent unique assemblages of uncertain affinity. Multidimensional analysis support the recognition of the groups with no overlap (fig. 5.7B). In support of the group formation, similarity analysis indicated that significant differences in composition and relative abundance among assemblage groups existed ($R = 0.976$, $P < 0.001$).

Agglomerative cluster and ordination analysis suggest appreciable changes in species composition through time. Visual inspection of the Dimension 1 assemblage scores plotted in stratigraphic order (i.e., from oldest to youngest) indicates that coral assemblages were subject to dramatic species changes from the Late Miocene to Middle to Late Pliocene, but the following Late Pliocene assemblages experienced continuous but minor changes in species composition ($R^2 = 0.74$, $n = 40$; fig. 5.8). Results from SIMPER analysis reveal the major species responsible for differences among assemblage groups. Group A1 is mainly defined by the contribution of *P. carrizensis, D. merriami*, and *S. fairbanksi*; group A2 is distinguished by the contribution of *S. fairbanksi*; group B1 is distinguished by the importance of *P. panamensis*; group B2 is mainly distinguished by the importance of *P. panamensis* followed by *P. capitata* and *P. damicornis*; finally, group B3 is distinguished by the equal contributions of *P. capitata, P. damicornis, Pocillopora verrucosa*, and *P. panamensis* and the relatively minor but equal importance of *Pocillopora meandrina, P. gigantea*, and *P. stellata*. Thus, there appears to be change in species composition, as well as the relative abundance of species in the assemblages, from Late Miocene to Recent.

Evolutionary History of the Gulf of California Coral Fauna

The coral species recognized in the Gulf of California result in increases of 118 percent in the total known Late Miocene to Recent fauna, and except for *Eusmilia carrizensis* from the Late Miocene Imperial Formation (Vaughan 1917) and *Pocillopora guadalupensis* from the Upper Pleistocene of Isla Guadalupe (Durham 1980), it comprises the entire Late Miocene to Late Pleistocene hermatypic fossil record in the eastern Pacific (López-Pérez 2005). Nevertheless, the species-area curve (fig. 5.4) suggests it is possible that not all species in the gulf region were sampled. The relatively low number of Late Miocene to Pliocene assemblages (table 5.1) and the shape of Late Miocene to Middle to Late Pliocene species-area curves (fig.

5.4C) indicate that Miocene to Pliocene strata, rather than the Pleistocene strata, are most likely to yield additional species. It is highly probable that the relatively low number of assemblages and species collected from the Miocene and Pliocene preclude the leveling off of the species-area curve, because the curve does not rise rapidly in the early stages of species accumulation. In contrast, the number of Pleistocene assemblages (table 5.1), the shape of the Pleistocene species-area curve (fig. 4C), and the fact that except for the rare and relatively recently recorded *Psammocora haimea, Psammocora brighami, Fungia curvata, Fungia distorta*, and *Fungia vaughani* (Reyes-Bonilla and López-Pérez 1998; Reyes-Bonilla 2002), all living Gulf of California species have fossil representatives. This reinforces expectations that better sampling of previously recorded localities and new exposures from a wider range of environmental settings from Miocene and Pliocene strata will render a larger number of species.

Previous studies of the Neogene history of reef corals in the Gulf of California supported a faunistic relationship with the Caribbean region (Vaughan 1917; Durham and Allison 1960; Dana 1975; Heck and McCoy 1978; Cortés 1986; Glynn and Wellington 1983; Budd 1989; Reyes-Bonilla 1992). Indeed, earlier Caribbean analyses considered the former as part of a much more extended pre–Central American Isthmus Caribbean region (Frost 1977a, 1977b; Budd and Johnson 1999a). Previous (Foster 1979, 1980; Budd 1989, 1991) and ongoing morphologic species comparisons suggest that, although closely related, Gulf of California species are sufficiently distinct to be considered as different. Except for *D. eminens, D. sarasotana*, and *S. mendenhalli* also occurring throughout Pliocene strata in the Caribbean (Budd et al. 1994), it may be suggested that the rest of the species are regional endemics. Based on available data, pre-turnover species originated between the Late Miocene (Imperial Valley) and Middle to Late Pliocene (Loreto area) in the Gulf of California. However, improved sampling and age resolution of Miocene deposits may extend the range of many species downward, thus smoothing the apparent Late Miocene–Early Pliocene peak in species origination. For instance, at Isla Tiburón debatable K-Ar ages of volcanic rocks overlying fossiliferous marine sediments containing *S. fairbanksi* and strontium isotope dates on calcareous megafossils indicate an age of 13–12 Ma (Gastil et al. 1999), which would have a profound influence on turnover metrics. Conservative estimates suggest a modest origination

Table 5.2. List of species identified in collections and arranged by taxonomic order.

Family	Genus	Species	No. of specimens	No. of assemblages
Pocilloporidae	Pocillopora	capitata	111	12
Pocilloporidae	Pocillopora	damicornis	6	4
Pocilloporidae	Pocillopora	meandrina	3	1
Pocilloporidae	Pocillopora	elegans	4	2
Pocilloporidae	Pocillopora	verrucosa	1	1
Agariciidae	Pavona	clavus	94	6
Agariciidae	Pavona	gigantea	4	2
Agariciidae	Gardineroseris	planulata	22	1
Siderastreidae	Siderastrea	mendenhalli	10	3
Siderastreidae	Siderastrea	sp.A	14	1
Siderastreidae	Psammocora	stellata	16	4
Poritidae	Porites	panamensis	237	28
Poritidae	Porites	carrizensis	99	8
Poritidae	Porites	lobata	1	1
Poritidae	Porites	sverdrupi	16	2
Faviidae	Favia	sp.A	39	1
Faviidae	Favia	sp.B	7	1
Faviidae	Diploria	bowersi	4	3
Faviidae	Diploria	sarasotana	2	1
Montlivaltiidae	Placosmilia?	sp.A	19	3
Faviidae	Solenastrea	fairbanksi	182	8
Meandrinidae	Dichocoenia	merriami	67	5

Total no. genera = 11; total no. species = 23; total no. specimens = 988.

rate during the entire Pliocene, although nonconservative estimates indicate the rate increase was limited from the Early to Middle Pliocene, when the last pre-turnover species were added to the gulf record (table 5.2).

Considering a Late Miocene (approximately 6.5 Ma) and Early to Middle Pliocene (5 Ma) age for the pre-turnover fauna, these taxa dominated gulf reef systems for approximately 3–3.5 Ma before their extinction during the Pliocene–Pleistocene transition. Although a more intensive sampling scheme and better resolution of age dates for the Late Miocene to Pliocene are necessary to unambiguously determined whether the originations were clustered (nonconservative estimates) or spread out (conservative estimates) in time, the data presented here indicate a relatively substantial number of originations in pre-turnover species followed by an extinction episode.

Earlier accounts of eastern Pacific coral reef evolution suggested that at some time between the Miocene and Pliocene a sharp differentiation between Caribbean and eastern Pacific hermatypic coral faunas took place (Vaughan 1917), although more recent accounts indicate the change might have occurred during the Pliocene–Pleistocene transition (Dana 1975; Glynn and Wellington 1983; Cortés 1986; Budd 1989; Reyes-Bonilla 1992). Conservative estimates based on new data suggest that the gulf fauna experienced constant extinction rates during the Pliocene with relatively minor increases during the Middle to Late Pliocene, whereas nonconservative metrics imply that the fauna underwent two severe extinction pulses concentrated at 3 and 1.5 Ma. Conservative and nonconservative estimates agree, however, that the lowest extinction rate occurred immediately after the Pliocene–Pleistocene transition. This shows that regardless of extinction

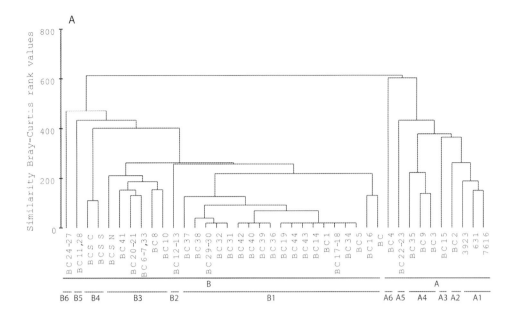

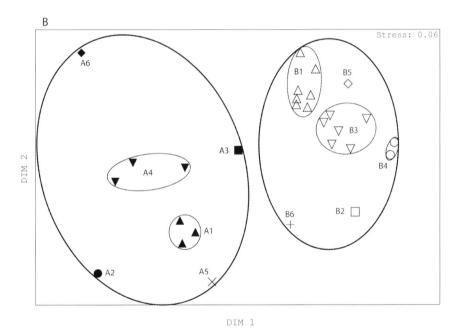

Figure 5.7. Classification of Late Miocene to Recent coral assemblages in the Gulf of California, Mexico. Dissimilarities among assemblages were generated from a Bray-Curtis similarity matrix. (A) Dendrogram showing results of a cluster analysis of assemblages; linkages were based on weighted pair group averages; and (B) distance map of the same assemblages produced by multidimensional scaling.

scenario, pre-turnover species do not pass through the age boundary. More accurate age dates are necessary to unambiguously determine whether extinction was spread over the Middle to Late Pliocene or was concentrated in the two extinction pulses. Relatively major losses in species richness directly after the extinction pulses, however, suggest that the pulse extinction scenario is more likely. During the extinction episode, except for *P. panamensis* and *P. capitata*, all

pre-turnover species were eliminated, leaving no living representatives.

Turnover metrics, especially those figures related to high origination rates during the Late Miocene and Late Pleistocene, may be misleading and its literal acceptance is not advised. The Late Miocene origination rate, for example, resulted because these are the oldest samples in the analysis, although they may not represent the oldest record of the species in the area, and

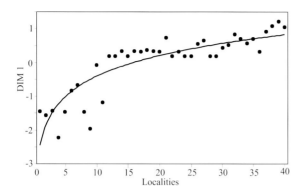

Figure 5.8. Relationship between Late Miocene to Recent coral assemblage scores (DIM 1) and time at the Gulf of California. Curve was constructed by adding assemblages in stratigraphic order. Each point represents an assemblage. Continuous line represents the best fit.

further age changes may smooth the trend. Similarly, Late Pleistocene origination rates probably resulted from Recent trans-Pacific colonization and do not necessarily reflect a presence any time before the Recent (i.e., Early to Late Pleistocene).

Faunal turnover (i.e., origination and extinction) had a profound influence on generic-level composition of pre- and post-turnover taxa. The opening of the gulf disrupted the species origination within genera that existed or that currently is restricted to the Atlantic/Caribbean, such as *Diploria*, *Solenastrea*, *Dichocoenia*, *Placosmilia*, *Favia*, and *Siderastrea*. Although species of *Favia* and *Siderastrea* also have been recorded from Recent Indo-Pacific and eastern Pacific reefs, recent phylogenetic analysis indicates that the Indo-Pacific *Favia* does not belong to the Atlantic/Caribbean clade (Budd and Johnson 1999a; Fukami et al. 2004), whereas eastern Pacific *Siderastrea glynni* is genetically indistinguishable from *Siderastraea siderea* (Forsman 2003). These figures reinforce previous statements suggesting high affinity between Caribbean and eastern Pacific faunas (Vaughan 1917; Durham and Allison 1960; Durham 1966; Glynn and Wellington 1983; Cortés 1986; Budd 1989; Colgan 1990; Reyes-Bonilla 1992). The extinction events that occurred in the area eliminated Gulf of California endemics and deprived the eastern Pacific of Atlantic/Caribbean genera. Following the extinction, post-turnover origination resulted from the immigration of species formerly included in circumtropical genera but that relatively long ago became restricted (except *Porites*) to the Indo-Pacific area. Species included in the genera *Pocillopora*, *Pavona*, *Gardineroseris*, and

Psammocora were commonly distributed in Caribbean areas, but those species vanished between the Early Miocene and Late Pleistocene (Geister 1975; Budd et al. 1994, 1996; Budd 2000).

Between peaks of pre-turnover species origination and extinction, extant *P. panamensis* and *P. capitata* were added to the Gulf of California. Between about 5 and 1.8 Ma, both taxa occur in assemblages of variable combinations of extinct and living species, suggesting that extinct and living species co-occurred not only among but also within assemblages. The oldest record of *P. panamensis* is in the Lower Pliocene San Marcos Formation at Ensenada El Muerto in the Punta Chivato area and at Puerto de la Lancha on Isla Carmen, where relatively small but abundant massive and ramose colonies contributed to the community structure. In the Middle to Upper Pliocene undifferentiated Carmen/Marquer Formation on Isla Monserrat, small colonies of *P. panamensis* co-occur with *P. carrizensis*, *Favia* sp. A, and the relatively abundant *P. capitata*. Multidimensional and SIMPER analyses do not support the role of biological interactions (i.e., ecological replacement) as the primary factor in the extinction of pre-turnover species. Instead, living species became important immediately after the demise of the former species.

Both conservative and nonconservative estimates agree that, except for *P. panamensis* and *P. capitata*, surviving species started to increase in number and relative abundance during the Pleistocene immediately after the demise of pre-turnover species. Prior to the Pliocene–Pleistocene transition, except for the recorded *Pocillopora* sp., *Leptoseris* sp., and *Porites* sp. in the Late Miocene–Early Pliocene Nazca Ridge region of the southeastern Pacific (Allison et al. 1967), there are no unambiguous data suggesting either successful or failed long-distance Indo-Pacific colonization of the Gulf of California or anywhere else in the eastern Pacific. Data suggest that after the arrival of *P. clavus* at La Ventana near the mouth of the gulf during the Early Pleistocene, transpacific colonization became a common and relatively successful event. This goes far to explain the high Pleistocene origination rate for coral reef evolution in the gulf. After addition of *P. clavus* to the gulf fauna, there was a steady arrival of Indo-Pacific species to the area.

Although relatively widely distributed between La Ventana and Isla Coronados, the fauna's relative abundance was meaningless because *P. panamensis* was the main coral reef builder during the entire Pleistocene. By the Early Pliocene, abundant massive and ramose colonies of *P. panamensis* contributed to the community structure at Punta Chivato, and from the Early to Late Pleistocene they formed monospecific reefs and constructions of variable combinations with relatively few species, yet always dominated by *P. panamensis*. From the Middle to Late Pleistocene, the relative importance of Indo-Pacific immigration increased, but not until the Recent were taxa other than *P. panamensis* able to contribute in approximately the same amount to the reef construction in Gulf of California localities (Reyes-Bonilla and López-Pérez 1998; Reyes-Bonilla et al. 2002) and to overcome *P. panamensis* elsewhere in the Mexican Pacific (Reyes-Bonilla et al. 2002; Reyes-Bonilla 2003; López-Pérez and Hernández-Ballesteros 2004) and eastern Pacific (Glynn and Wellington 1983; Glynn and Ault 2000). Qualitative comparisons between living and Pleistocene assemblages show that *G. planulata* and *P. lobata* failed to successfully colonize the gulf. *Gardineroseris planulata* formed relatively large and numerous colonies during the Middle Pleistocene at La Ventana, but disappeared afterward. No positive record of the species in the Gulf of California or neighboring areas has been established; instead, the closest definite record of the species is in the Bahías de Huatulco area of the Mexican Pacific (Leyte-Morales 1995). *Porites lobata* is a widely distributed Indo-Pacific species recently recorded at Bahía de Banderas, Nayarit, in the Mexican Pacific (Reyes-Bonilla et al. 1999), but the species reached Cabo Pulmo near the mouth of the gulf during the Late Pleistocene, and similar to *G. planulata* it disappeared afterward from the area. In summary, after the demise of pre-turnover taxa during the Late Pliocene, the long-distance dispersal, arrival, settlement, and development of Indo-Pacific species in the area was a likely and successful outcome in the evolution of the Gulf of California coral-reef fauna.

Taxonomic turnover metrics are similar to those observed in the Caribbean area. In particular, the extinction rates at 2–1 Ma were also a key factor in the richness, composition, and relative abundance of species in the Recent Caribbean reef systems (Budd et al. 1996; Budd and Johnson 1997, 1999b; Budd 2000; Johnson 2001). The observed turnover is, nonetheless, unique in that: (1) the regional origination of pre-turnover species probably resulted from the formation of a geologically and biogeographically critical site at the northern boundary of the eastern Pacific

tropics; and (2) instead of reducing species richness as occurred in the Caribbean (Budd et al. 1996; Budd and Johnson 1997, 1999b; Budd 2000; Johnson 2001), the extinction event triggered the long-distance Indo-Pacific colonization of species.

There has been significant improvement in age resolution for Gulf of California outcrops (table 5.1), but more accurate age dates are necessary to establish a causal relationship between environmental changes and their relative importance in gulf turnover rates (i.e., origination and extinction). The extinction scenario has received special attention in discussing the evolution of eastern Pacific coral fauna. Since the possible devastating role of the Pleistocene environmental perturbations (especially shallower thermoclines, reduced salinities, and increased turbidities) to reef growth in the eastern Pacific was mentioned by Dana (1975), a relatively large number of studies supported the Pleistocene extinction scenario (Glynn and Wellington 1983; Cortés 1986; Colgan 1990; Reyes-Bonilla 1992). Nonetheless, turnover metrics indicate that extinction predated Pleistocene environmental perturbations and, thus, was probably more closely tied to environmental perturbations that resulted from the closing of the Isthmus of Panama. However, except for the origination scenario that focuses on the trans-Pacific immigration of Indo-Pacific species (Glynn and Wellington 1983), there is no account regarding the origination of pre-turnover species. In both cases, the common denominator is related to the opening of spaces to coral immigration, settlement, and development. It is likely, for example, that pre-turnover species arose after the geographic differentiation of northwestern populations of widely distributed Caribbean species.

To better understand the evolutionary history of the Gulf of California coral fauna, fundamental questions still remain, including the cause of faunal turnover, the relative importance of biological and environmental factors in the faunal change, and the evolutionary histories of *P. panamensis* and *P. capitata* and their relationship with Caribbean and Indo-Pacific taxa. In addition, although new data allow us to improve our understanding of the evolution of the Gulf of California coral fauna, more samples and more accurate age resolution of the coral-bearing outcrops are needed to confirm the main evolutionary trends.

6 Living Rhodolith Bed Ecosystems in the Gulf of California

Diana L. Steller, Rafael Riosmena-Rodríguez, and Michael S. Foster

Rhodoliths are unattached, rose- to lavender-colored, spherical, branching algae that resemble coral and that rock and roll around on the seafloor. Technically, they are nongeniculate coralline red algae that may also grow as crusts on rock (Corallinales, Rhodophyta), photosynthesize, and are impregnated with calcium carbonate. Populations of living individuals are often stacked upon each other, and the carbonate sediment they produce forms a biogenic habitat called a rhodolith bed (fig. 6.1A–C). These beds are distributed worldwide (Foster 2001), and the complex habitat they create supports a rich and diverse community of other algae and animals over soft-sediment benthos (Grall and Glemarec 1997; Hall-Spencer 1998; Steller et al. 2003; Foster et al. 2007). Unlike most fleshy algae, the carbonate thalli of whole or fragmented rhodoliths are often preserved in fossil deposits, providing ecological data on past shallow-water environments (see chap. 7). Despite their widespread distribution and in contrast to most other common marine habitats, the basic ecology of rhodolith beds is poorly understood.

Rhodoliths in the Gulf of California have been known to scientists since the late nineteenth century (Hariot 1895). However, it was not until the early 1990s that the exceptional abundance of fossil and living beds, their high diversity, their importance to near-shore ecology, and the need for their conservation in the gulf began to be appreciated (reviewed in Foster et al. 1997). The global distribution of rhodoliths and an increased awareness of their importance and susceptibility to disturbance have led to an increased appreciation of rhodolith habitats worldwide. In this chapter we review the current knowledge of living rhodoliths in the Gulf of California and discuss this information relative to their present and future conservation.

History of Rhodolith Exploration in the Gulf of California

Leon Diguet, a French naturalist from the Natural History Museum of Paris, was in charge of the evaluation of natural resources along the Baja California peninsula (Diguet 1895) and participated in several gulf expeditions between 1893 and 1896 (Woelkerling and Lamy 1998). His observations were related to the extraction of pearls (Diguet 1899) and evaluation of other resources such as marine mammals (Diguet 1911). Diguet considered the development of the pearl-oyster culture by Gaston Vives to be the most important resource-use developed during his experience along the coast of La Paz (Diguet 1899, 1911).

During the late nineteenth century, the economic and cultural influence of France in Mexico was clearly reflected by the Baja California fisheries culture (Cariño-Olivera 1995). Gaston Vives' pearl culture method, the first known in Latin America (Cariño-Olivera 2000), spanned ocean larval collections to adult growth and pearl development (Cariño-Olivera and Monteforte 1999). In the final stages of this procedure, divers prepared areas for the adult animals to develop using *chicharrones* as the benthic substrate (Cariño-Olivera and Caceres-Martinez 1990). The name *chicharrones* (meaning "dry pork skin") is used by fishermen to describe the natural rhodolith habitat where pearl oysters and other marine shellfish resources were found.

Diguet's algal collections from the La Paz region were derived mostly from coralline algal "chicharrones material," the taxonomy of which was first worked on by Hariot (1895; see also Woelkerling and Lamy 1998) and then other European specialists (reviewed in Riosmena-Rodríguez et al., in press). As a result of extensive fieldwork between 1940 and 1960, E. Y. Dawson (1944) made the first comprehensive

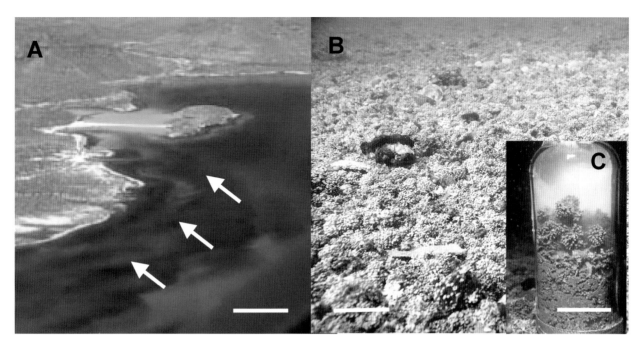

Figure. 6.1. A rhodolith bed near Isla Requesón in Bahía Concepción: (A) aerial photograph (arrows indicate bed; scale = 500 m); (B) underwater view of the bed surface (scale = 10 cm for foreground and 1 m for background); (C) clear core through the surface showing living rhodoliths grading into sediment with rhodolith fragments (scale = 5 cm). Photo credits: (A) Michael S. Foster; (B, C) Diana L. Steller

analysis of the Gulf of California coralline flora (Dawson 1960). At that time, at least 15 specific names from two genera (*Lithothamnion* and *Lithophyllum*) were used for the gulf rhodolith material. A later taxonomic change made by Adey and Lebednick (1967) added a new genus with *Neogoniolithon trichotomum*. A growing appreciation of the great morphological variability led Riosmena-Rodríguez et al. (1999) to reduce the number of described *Lithophyllum* species from five to one: *Lithophyllum margaritae*. Present taxonomic evidence indicates that only four living rhodolith species from four genera occur in the Gulf of California (review in Riosmena-Rodríguez et al., in press): *L. margaritae*, *Lithothamnion muelleri* (*Lithothamnion crassiusculum*), *N. trichotomum*, and *Mesophyllum engelhartii*. There may be a fifth species, depending on the pending taxonomic resolution of *Lithothamnion australe* (Riosmena-Rodríguez 2002).

Spatial Distribution

Living rhodolith beds have been found throughout the coasts of the Gulf of California from Puerto Peñasco, Sonora, near the Colorado River delta to the south off Islas Marietas, Jalisco, and likely beyond (fig. 6.2; note: specific bed numbers throughout the chapter refer to this figure). The distribution is not completely mapped, as beds are often not visible from the ocean surface, commonly occur in areas thought to be simply soft bottoms, and can shift in location. In the Gulf of California, living rhodolith beds consistently have been found at semiprotected sites around islands and in bays and channels. Intensive surveys have only been conducted in the regions around Mulegé and La Paz (fig. 6.2B, D). The shoreline extent and depth of most of these beds are still unknown. The higher bed densities along the peninsular gulf coast presented here may result from oceanographic variation, lower sedimentation, dominant wind/water motion direction, or simply increased search time.

Determining the current distribution of living beds from areas of dead rhodolith or other carbonate sediment requires a variety of approaches, including site descriptions from previous live collections (Diguet 1895; Dawson 1960), knowledge from local fisherman, and a combination of satellite images, aerial photos, dredging, and diving surveys. Banks of carbonate sediment usually dominate geographic areas where living beds exist. These are often visible above the water as carbonate beaches (e.g., the tomobolo at El Requesón, bed no. 27) and dunes (Sewell et al. 2007) or submerged as dark patches against a white background.

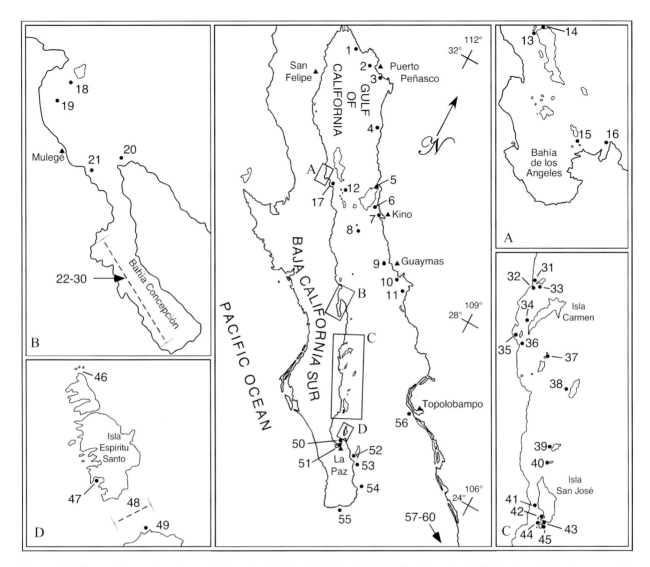

Figure 6.2. Maps of known locations of living rhodolith beds in the Gulf of California; a rhodolith bed consists of an area of 10 percent or greater cover of rhodoliths. All beds except where designated EYD (Dawson 1960) or O (other) were observed by one or more of the authors; Dawson's observations were based on dredged material or not specified. Numbers correspond to the following sites by region. *Puerto Peñasco/Guaymas Region*: (1) Punta Borrascosa (O); (2) Manto Peñasco (O); (3) Estero Morua (O); (4) Cabo Tepoca; (5) Estero Arenas (O; (6) Estero Santa Rosa; (7) Isla Alcatraz; (8) Isla San Pedro Mártir (O); (9) Isla San Pedro de Nolasco; (10) Las Gringas; (11) Bahía Bocochibampo (EYD); *Bahía de Los Angeles Region*: (12) Isla Partida (EYD); (13) Puerto Refugio (O); (14) Canal Mejia (O); (15) Isla Cabeza de Caballo; (16) Bahía de San Francisco Pond Island (EYD, O); (17) Bahía de Las Palomas (O); (18) Punta Chivato I; (19) Punta Chivato II; *Bahía Concepción Region*: (20) Punta Aguja (Concepción, EYD); (21) Los Machos; (22) Santispac; (23) Isla Blanca; (24) Isla Coyote; (25) Morro Tecomates; (26) El Cardón; (27) Isla El Requesón; (28) La Cueva (Correcaminos); (29) Los Pocitos; (30) El Coloradito; *Loreto Region*: (31) Punta Bajo; (32) Isla Coronados Channel; (33) Isla Coronados; (34) Isla Carmen (in Bahía Salinas; EYD, O); (35) Isla Danzante (O); (36) Puerto Escondido (EYD); (37) Islas Las Galleras; (38) Isla Catalina (O); (39) Isla Santa Cruz (EYD); (40) Isla San Diego; (41) Isla San José; (42) Isla San José Estero; (43) Isla El Pardito; (44) La Lobera; (45) Isla San Francisquito; *La Paz Region*: (46) Los Islotes (O); (47) Bahía San Gabriel; (48) Canal de San Lorenzo; (49) Punta Galeras; (50) Isla Gaviota; (51) Bahía de La Paz (Malecon/historic site; EYD, O); (52) Isla Cerralvo (O); (53) Punta Perico; (54) Cabo Pulmo (O); (55) Bajo La Gorda (EYD); *Sinaloa Region*: (56) Topolobampo (EYD); (57) Isla Venado; *Nayarit Region*: (58) Isla Isabel (O); (59) Isla María (O); (60) Isla Marietas. Data from Riosmena-Rodríguez et al. (in press).

Using this contrast, shallow beds are often visible on aerial photographs (fig. 6.1A) and satellite images (fig. 6.3A). However, satellite images may not resolve distributions of deep or low-density beds. Known beds in the Isla San José region are not (La Lobera, bed no. 44) or are not entirely (Isla El Pardito, no. 43) detectable in figure 6.3A. While satellite images clearly enhance detection of large-scale, shallow carbonate deposits, variation in water clarity (generally about 10–15 m) may obscure bed locations even in shallow water (< 15 m), limiting the utility of this tool for bed detection. The presence of a living bed can only be verified using diving surveys or dragging/coring and immediate sample sorting to verify the presence of

live rhodolith material. Steller and Foster (1995) used a combination of aerial photographs and diving surveys, whereas Hetzinger et al. (2006) used acoustic surveys combined with sediment cores.

Other carbonate-producing organisms are important throughout the gulf (Halfar et al. 2004) and can be found in rhodolith-rich areas. In a carbonate-rich area around southern Isla San José (fig. 6.3B; bed nos. 42–45), Hetzinger et al. (2006) found that, in addition to the approximately 40 percent of the bottom of a 45-km² area that was rhodolith-derived material, bivalves, bryozoans, and corals were also important carbonate contributors. However, core material from the study was not differentiated into living relative to

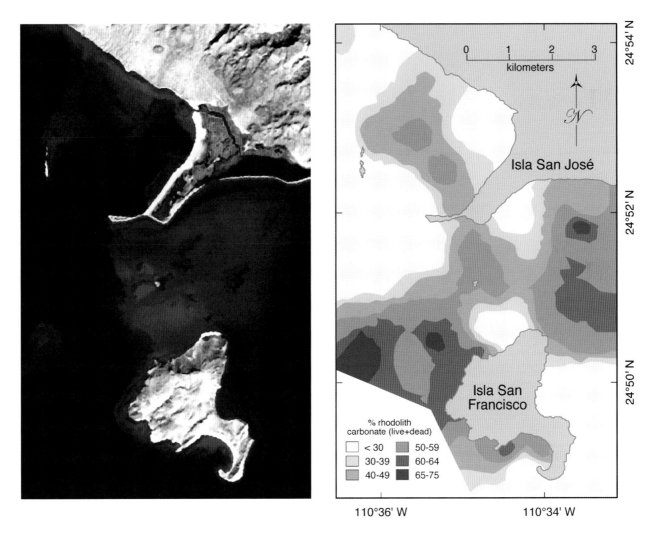

Figure 6.3. Comparison of rhodolith distribution patterns from different sources: (A) satellite image of southern end of Isla San José derived from an Enhanced Thematic Mapper satellite image using bands 4,2,1 modified using Gram-Schmidt spectral sharpening; and (B) percent cover of rhodolith-derived carbonate material (live + dead) based on Hetzinger et al. (2006). ETM+ image acquired from University of Maryland Global Land Coverage Facility. The original data set was collected by the NASA LANDSAT Program, LANDSAT ETM+ scene L7CPF20011001_20011231_05, L1G, USGS, Sioux Falls, October 26, 2001.

dead rhodolith material, thereby losing valuable information about living source populations. Although the map they produced from core surveys does not differentiate live and dead material (fig. 6.3B), the areas of high concentration reflect the same kind of living populations described in previous studies by Foster et al. (1997). The identification of live material in future coring studies and the combination of this information with acoustic surveys may help clarify the extent of living rhodolith beds.

Physical Setting

Like all benthic algae, living rhodoliths only occur where light on the bottom is suitable for growth. Assuming a minimum light requirement of around 1 percent of surface illumination, Foster et al. (1997) estimated that this minimum occurs at a depth of about 30 m in the gulf. This may be an underestimate, however, if the minimum light requirements of gulf rhodolith species are lower and water at a particular site is more transparent. Living rhodoliths have been found to depths of about 150 m in other regions (Harris et al. 1996). The deepest living beds so far found in the gulf are at 25–40 m (e.g., a current bed at Punta Perico, bed no. 53; Vázquez-Elizondo 2005; Riosmena-Rodríguez et al., in press). While more information is needed to accurately understand light requirements, especially as they may account for among-species distributional differences, it appears that living rhodolith beds in the gulf may be restricted by light to shallow waters along the Baja California peninsula and its associated islands.

Bed distribution is further constrained by suitable combinations of bottom topography and sedimentation, the latter a function of water movement and sediment input from coastal runoff. These factors interact, obviating simple cause-and-effect relationships. Rhodolith annual growth rates (0.6–4.9 mm/year in radius; reviewed in Steller 2003; Riosmena-Rodríguez et al., in press) and common maximum sizes (3- to 8-cm diameter; Foster et al. 1997; Steller 2003; Rivera et al. 2004) in the Gulf of California indicate that individuals are retained in a bed at least on the order of years to tens of years, assuming little to no exchange among beds. Thus, beds are found on flat or gently sloping bottoms where these unattached algae can remain and grow for long periods of time. Topography also affects water motion, depending on orientation to prevailing waves and as influenced by tidal currents. Waves and tidal currents are often oscillatory with

periods of seconds to a few hours, contributing to the retention of rhodoliths as well as unattached forms of other organisms, such as corals and bryozoans, occurring within rhodolith beds (Reyes-Bonilla et al. 1997; James et al. 2006). Transport may not be deleterious in areas such as the western shore of Bahía Concepción (bed nos. 23–30), where waves coming ashore at an angle might cause net movement from one favorable habitat to another, or in areas such as northern Bahía Santa Inés (Punta Cheviot, bed nos. 18 and 19), where unidirectional currents may simply move rhodoliths around over a large area of favorable habitat.

Rhodolith beds generally occur in areas of moderate water motion (Bosence 1983; Foster et al. 1997). Movement induced by rough-water motion can fragment rhodoliths or transport them to unfavorable habitats. Moderate water motion has been suggested as necessary for bed persistence, as it may turn individual rhodoliths and thereby facilitate growth in all directions, inhibit fusion of thalli into larger crusts, reduce fouling, and/or inhibit sediment accumulation on thalli. While all of these relationships are possible, turning can also result from bioturbation (Bosence and Pedley 1982; Marrack 1999), and only the effects of sediment accumulation have been clearly demonstrated. Steller and Foster (1995) found that thalli transplanted below the lower limits of a rhodolith bed were buried by fine sediment and died. Wilson et al. (2004) showed that anoxia associated with fine sedimentation can kill thalli in two weeks in the laboratory, and Hall-Spencer et al. (2006) found that rhodoliths in the field were killed by anoxic conditions caused by deposition of organically enriched, fine sediments. Therefore, rhodoliths seem to be very sensitive to anoxia, and moderate water motion may be most important to bed persistence as it inhibits the settlement of fine sediments or reduces their accumulation via resuspension. The former process is likely to be more common with currents, and the latter with waves. Water motion also transports planktonic larvae and food into beds and may directly contribute to associated high faunal abundances. The fauna, in turn, can increase local nutrient concentrations (Martin et al. 2007) and bioturbation. Variation in nutrients in temperate beds is considered not to be limiting (Martin et al. 2007), but it may affect rhodolith species composition (reviewed in Bosence 1983). Rhodoliths appear to be tolerant of all but extreme variations in salinity (Wilson et al. 2004). In the Gulf of California, rhodoliths tolerate the extreme temperature fluctuations found in the gulf (8–32°C; Alvarez-

Borrego 1983). However, significantly different seasonal growth rates (Rivera et al. 2004; Steller et al. 2007a) and reduced photosynthetic capabilities measured at low and very high temperatures suggest that further studies are required to understand the overall effects of temperature on rhodolith production.

Field observations in the gulf have shown that living rhodolith beds occur in two primary water-motion regimes, wave-dominated and current-dominated, and beds have been characterized accordingly as wave beds and current beds (Foster et al. 1997). Wave beds are generally large (several square kilometers), and dense rhodolith concentrations are found along sloping platforms from 2–12 m deep in areas where water motion occurs primarily from short-period waves produced by wind blowing over short fetches (fig. 6.4A, e.g., beds within Bahía Concepción, El Requesón bed no. 24). The resulting moderate, oscillating water movement (maximum of 38 cm/sec, bed no. 29, as measured by Marrack 1999) at a depth of 5–6 m in Bahía Concepción resuspends fine sediments such that accumulation is reduced. At depths above 2–5 m, higher water velocities cause fragmentation and dispersal (Steller and Foster 1995). Thus, these live pigmented wave beds often show distinct boundaries at their shallow edges with coarse rhodolith fragments and at their deeper edges with fine sediments. The rhodolith-forming species *L. margaritae* is common and *L. muelleri* may be present throughout these wave beds, and the more fragile species *N. trichotomum* can be abundant in shallow-water beds in the southern gulf (Riosmena-Rodríguez et al., in press).

Populations in wave beds can be dense (up to about 10,000 per m² as verified by Steller et al. 2003) and are often dominated by small individuals with a few large ones (to about 4- to 5-cm diameter). Rhodolith morphology often varies from larger, densely branched to smaller more sparsely branched, likely a result of variation of multiple physical factors (Steller and Foster 1995) and a feature that surely affects the potential for fossilization. Higher energy wave beds (fig 6.4B) have also been found between 1- to 7-m depths in areas of mixed boulders and coarse sand partially exposed to waves generated in the gulf (e.g., Cabo Los Machos, Punta Galeras, bed nos. 21, 49; Foster et al. 1997, 2007). The rhodoliths in these beds are a lumpy form (thick, densely packed cylindrical branches) of *L. muelleri*, and they occur as scattered large individuals. *Sargassum* spp. can be abundantly attached to the boulders in the area. The upper limits of these beds may be determined by exposure to air during low

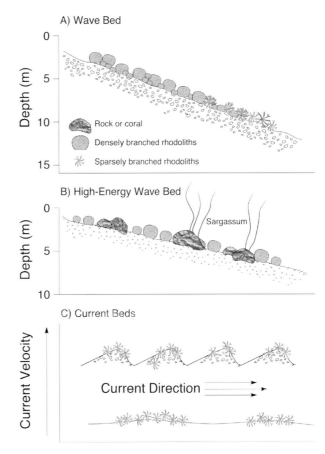

Figure 6.4. Idealized cross sections through typical living rhodolith beds in the Gulf of California relative to dominant water motion: (A) wave bed: high-density rhodoliths; rhodolith morphology variable relative to depth; distinct depth boundaries; bed underlain by carbonate of mixed grain size plus fine siliciclastic sediment; common species *Lithophyllum margaritae*, *Neogoniolithon trichotomum*, and *Lithothamnion muelleri*; example Isla El Requesón, bed no. 24; (B) high-energy wave bed: reduced depth distribution with gentle slope; patchy, low-density large rhodoliths; rhodoliths densely branched throughout; rhodoliths on medium-grain sediment between corals or boulders; seasonal *Sargassum horridum* on boulders; example Los Machos, bed no. 21; (C) current bed under variable current speeds: patchy; rhodoliths underlain by pure carbonate sediment; low density overall with high-density patches at low velocity; formation of ripples with rhodoliths on ridges at higher velocities; example San Lorenzo Channel, bed no. 48.

tides, but the causes of the lower limits are unknown. Water velocities within the beds during large wave events have not been measured.

Current beds are generally larger than wave beds (up to 10 km²) and are most commonly found in areas where water movement is caused by strong tidal currents (up to 37 cm/sec at 12–14 m deep in Canal de San Lorenzo, bed no. 48, and Isla Coronados Channel, bed no. 33; Marrack 1999). Living rhodoliths often occur at low densities with a patchy distribution over carbonate sandy bottoms, forming ripples with rhodoliths on ripple crests when current velocities are high (fig. 6.4C). The deepest current bed found to date is 35–40 m at Punta Perico (bed no. 53; Vázquez-Elizondo 2005), and it is possible that the currents in this bed near the mouth of the gulf resulted from boundary currents like those described by Harris et al. (1996). Currents over rhodolith beds may also be generated by winds. Foster (personal observation) witnessed currents generated by strong westerly winds that were sufficient to turn and move rhodoliths at 7-m depth at Punta Chivato. Known current beds in the Gulf of California are dominated by *L. margaritae*, and beds have been found from less than 1 to 40 m deep, depending on the current distribution at particular sites. Accumulation of fine sediment has not been measured in current beds, but the lack of fine sediment in about 1-m-deep excavations in the center of beds in the Isla Coronados Channel and Canal de San Lorenzo (Foster, personal observation) suggests that currents may inhibit sedimentation. Fine sediment is more common at the margins of these generally large beds, suggesting that variation in water motion and sedimentation may determine their lateral and lower depth limits.

These conclusions concerning the abiotic environment of living beds are based primarily on observations during normal oceanographic conditions on an annual scale. The gulf, however, experiences frequent, episodic hurricanes and tropical storms, and the associated increases in water motion could have significant effects on rhodolith bed distribution or relocation and drive large-scale transport and deposition. Such effects are indicated by rhodolith-rich layers in deep-water (300 and 700 m) sediment cores (reviewed in Riosmena-Rodríguez et al., in press), abundant rhodoliths cast onshore after a hurricane (Foster et al. 2007), and temporal changes in current bed margins on the order of 30–40 m (Steller, personal observation). The structure (e.g., density and size-frequency distribution of rhodoliths, sediment composition) of a bed at any particular time may reflect disturbance and recovery from a hurricane in addition to the effects of normal, annual conditions. While the effects of hurricanes are difficult to determine, further knowledge of them seems essential to understanding the physical setting of living rhodolith beds and fossil deposits in the gulf.

These constraints based on local observations indicate that if there are larger-scale distribution patterns, they may result from large-scale differences in near-shore geomorphology, water motion, and sedimentation, a conclusion similar to that reached by Nelson (1988) in his review of nontropical shelf carbonates. The abundance of rhodolith beds may also be affected by variation in rhodolith growth rates due to temperature and calcium carbonate availability (Steller et al. 2007). The great variation in the occurrence and composition of rhodolith beds found within a local site (e.g., Bahía de La Paz; Foster et al. 1997; Halfar et al. 2000) argues for caution in attributing general causes based on sampling a few localized modern beds or fossil deposits, no matter how widely separated. Ecological studies have repeatedly verified the fundamental logic that until small-scale variation is understood, putative large-scale variation may be an illusion produced by inadequate sampling design (Underwood 1997).

The above conclusions are based on limited information concerning spatial distribution and largely on qualitative correlation. It is our hope that the increasing use of remote sensing to detect and map rhodolith beds (e.g., De Grave et al. 2000; Forrest et al. 2005), combined with traditional SCUBA sampling and continuous, bed-scale in situ recording of appropriate environmental variables (e.g., Mitchell and Collins 2004) will provide the descriptive information necessary to develop more rigorous hypotheses and that these will be tested by experiment.

Biological Setting

Members of a diverse community associated with Gulf of California rhodolith beds can live within underlying sediments (infauna); on, among, or bored into rhodolith branches (cryptofauna and -flora); or on top of rhodolith surfaces (macrofauna and -flora), or they may move among these microhabitats depending on life-history stage (Steller et al. 2003). In addition, transient species such as many fishes and large, mobile benthic invertebrates are periodically associated with beds in the Gulf of California. Examples include the highly mobile bay scallop (*Argopecten ventricosus*, fig. 6.5A), predatory gastropods (*Muricanthus nigritus, Hexaplex erythrostomus*) and sea stars (*Luidia* sp., *Astropecten* sp.), and sea urchins (*Toxopneustes roseus*,

Figure 6.5. Organisms commonly associated with rhodolith beds: (A) bay scallops (*Argopecten ventricosus*) in a rhodolith bed (scale = 3 cm); (B) common urchin (*Toxopneustes roseus*) carrying a rhodolith (scale = 5 cm); (C) a recruit of the brown alga (*Sargassum horridum*) on a rhodolith (scale = 1 cm). Photo credits: Diana L. Steller

fig. 6.5B). Most early studies presented here aimed to establish patterns of species association. Experimental work is necessary, however, to understand the mechanisms behind the complex ecological relationships.

Research on rhodolith bed diversity in the Gulf of California has revealed patterns of high abundance and species richness. In a study using stratified sampling of subhabitats in a rhodolith bed (El Coyote, bed no. 24) and a sand flat, Steller et al. (2003) reported a 1.7-times higher species richness and approximately 900-times higher abundance in a rhodolith bed when compared to the adjacent sand flat. The greatest contributors to higher richness and abundance were cryptofauna (primarily arthropods, annelids, and cnidarians), and their abundance was found to increase with rhodolith branching and size, a relationship important to conservation. This study demonstrated how stratified sampling, instead of using coring techniques to sample the community, provides valuable information about substrate associations.

Two intensive studies conducted on large rhodoliths examined patterns of cryptofaunal abundance and distribution in *L. margaritae* and *N. trichotomum* in the southwestern Gulf of California. A comparison of bed type (two wave and two current beds; bed nos. 33, 42, 43, and 48) and season (winter vs. summer) of cryptofauna in *L. margaritae* revealed 118 taxa in the 160 rhodoliths collected (Medina-Lopez 1999). There

were about 10 taxa and 32 individuals per rhodolith, and crustaceans were the most abundant taxa. Few differences between bed types or seasons were found, compared to the strong seasonal macrofaunal and -floral patterns reported by Steller et al. (2003) in the El Requesón bed (no. 27). Hinojosa-Arango and Riosmena-Rodríguez (2004) found a total of 104 cryptofaunal taxa (52 identified to species) in 120 large (4- to 5-cm diameter) individual rhodoliths (100 *L. margaritae* and 20 *N. trichotomum*) with similar branch densities, collected from two sites in Bahía Concepción and an Isla Coronados site (bed no. 33). Crustaceans were again the dominant group in all samples. Cryptofaunal diversity and abundance were similar in *L. margaritae* from the two sites in Bahía Concepción, averaging about 10 taxa and 75 individuals per rhodolith. Diversity and abundance were similar between the two rhodolith species from Isla Coronados, with the exception of higher crustacean abundance in *N. trichotomum*.

These studies suggest generally similar diversity and abundance among the cryptofauna in living rhodolith beds in the gulf dominated by *L. margaritae*/*N. trichotomum*, which have similar morphology. That rhodolith species with very different morphologies may harbor a different cryptofauna has been shown by recent surveys at Cabo Los Machos near the mouth of Bahía Concepción (Foster et al. 2007). The high-energy wave bed at this site (bed no. 21) is perennially dominated by abundant *L. muelleri* (10–50 individuals/m²) and seasonally by *Sargassum horridum*, which sometimes recruit to the rhodoliths (Fig. 6.5C), dominating the water column. The individual rhodoliths have very wide (about 1 cm), densely packed branches, such that the cryptofauna is composed primarily of boring organisms and other species that occupy spaces produced by borers. From 15 rhodoliths, a total of 117 taxa (107 species) were found, with polychaetes the most diverse group (49 taxa). Although from different sampling methods and times, data from all current surveys suggest that rhodolith cryptofaunal diversity is high and is similar among species with similar morphologies. An improved understanding of rhodolith bed structure and its variation in space and time would be facilitated by the use of similar sampling designs and methods as described in Steller et al. (2007b).

The species richness of rhodolith beds and associated sediment in the Gulf of California was further illustrated by the work of Cintra-Buenrostro et al. (2002). They found a total of 142 mollusk taxa larger than 1 cm (85 bivalves, 57 gastropods) in 1-×-0.2-m core samples taken from 10 living rhodolith beds in the Bahía Concepción/Punta Chivato region and the bed in Canal de San Lorenzo (bed no. 48). In addition to rhodoliths, other commonly attached organisms are often found living unattached in rhodolith beds. Reyes-Bonilla et al. (1997) found five species of free-living corals (coralliths) in rhodolith beds in the southwestern gulf, including several new distributional records. James et al. (2006) reported two bryolith (free-living bryozoan) species at sites in the central and southwestern gulf.

Considerable taxonomic work remains to be done on invertebrate collections from rhodolith beds in the Gulf of California and promises to reveal new species and biogeographic insights. For example, Clark (2000) identified eight species of small (2–10 mm) cryptofaunal chitons collected primarily from *L. margaritae* in the southwestern gulf, four of which were new to science. He commented that the chiton fauna of these rhodolith beds is "particularly rich and diverse." It is clear that rhodolith beds and particularly individual rhodoliths are diversity hot spots in the Gulf of California.

In addition to the rhodoliths themselves, rhodolith beds often harbor a diversity of other algae, and seasonal changes in oceanographic conditions in the Gulf of California typical of subtropical areas (Alvarez-Borrego 1983) appear to most strongly affect the diversity and cover of these associated organisms. Steller et al. (2003) found up to 30 associated macroalgal species per bed in four *L. margaritae* beds in the southwestern gulf in winter, but only up to eight species in summer. Nonrhodolith macroalgal cover was less than 2 percent in summer. Winter macroalgal diversity was higher in deeper beds compared to shallower beds, perhaps due to generally lower water temperatures, lower variation in water temperatures, reduced surge, and higher nutrient levels. Microalgal (diatoms and cyanobacteria) cover, however, was highest in summer. The high summer water temperatures, potential anoxic conditions at some sites, and thermal stratification likely influence low summer diversity and lead to the disappearance of most foliose red algae and the dominance of more tolerant microalgae (Lechuga-Deveze 2000). Similar seasonal patterns in macroalgal diversity have been described for rocky shores in the southern Gulf of California (Paul-Chavez and Riosmena-Rodríguez 2000) and in Bahía Concepción (Mateo-Cid et al. 1993).

Fish associations with rhodolith beds are not well known. Our qualitative observations suggest that fish diversity, at least of larger species that spend considerable time in the water column, is low. This is probably because at spatial scales larger than about 5–10 cm, most rhodolith beds are basically two-dimensional, providing very little structure to associate with. Adult fishes common in beds include species that feed on abundant benthic invertebrates, such as the Pacific porgy (*Calamus brachysomus*) and the spotted sand bass (*Paralabrax maculatofasciatus*). Reports of numerous associated fish (e.g., Alburto-Oropeza and Balart 2001) in rhodolith beds are likely an artifact of the beds being in close proximity to rocky or other reefs that provide macrostructure. Riosmena-Rodríguez (personal observation) has seen thousands of trigger fish (*cochi, Balistes polylepis*) spawning over the carbonate sediments produced by the rhodoliths in Canal de San Lorenzo. Thomson et al. (2000) also reported that trigger fish like to come to sandy habitats to feed. We have noted numerous gobies and blennies on, in, and among rhodoliths. These fishes may be quite diverse given the complex, small-scale structure of rhodolith beds. This would be an excellent topic for future study. Large, bottom-dwelling or burrowing species can be quite common, including tiger snake eels (*Myrichthys maculosus*), bullseye electric rays (*Diplobatis ommanata*) and other rays, and Cortez garden eels (*Taeniconger digueti*).

The distribution of rhodolith beds appears to be largely constrained by abiotic variables. However, within-bed community structure and perhaps even bed persistence may be strongly affected by species interactions. For example, bioturbation by invertebrates and fish may, by moving rhodoliths and resuspending fine sediment, facilitate rhodolith growth and bed maintenance (Marrack 1999; James et al. 2006). Nutrients produced by invertebrates may be important to rhodolith growth. The four new chiton species described by Clark (2000) may require the habitat provided by living rhodolith branches for their existence. Bivalves are abundant in rhodolith beds in the gulf and elsewhere (Hall-Spencer 1998, 1999; Hall-Spencer et al. 2003; Steller 2003). This can result from larval settlement preferences for coralline algae (Kamenos et al. 2004) or other structured or large-grain substrates, enhanced growth rates of species inside rhodolith beds, and possibly because the beds provide some refuge from predation (Steller 2003). These positive attributes of rhodolith beds, however, create a conservation dilemma because of the degradation that results from commercial fishing (Hall-Spencer 1998, 1999; Hall-Spencer et al. 2003). Algal species typically found in the diet of the green turtle (*Chelonia mydas*) and its foraging grounds (López-Mendilaharsu et al. 2005) in Bahía Magdalena are commonly associated with rhodolith beds or rhodolith sediment (López-Mendilaharsu et al. 2005). Interactions between biotic and abiotic factors may be especially important. For example, the water motion patterns responsible for rhodolith maintenance may also influence larval delivery of a number of associated species to beds. Such interactions remain to be investigated.

Rhodolith Conservation

Rhodoliths are dependent upon some level of disturbance by natural water motion. However, the anthropogenic disturbances of decreased water quality, extraction, and bottom fisheries (De Grave and Whitaker 1999; Hall-Spencer and Moore 2000) can contribute to the burial and destruction of thalli. Extreme levels of these disturbances can result in reduced thallus density, size, and structure; loss of living thalli; and ultimately transition into relatively low-diversity carbonate sand flats. The greatest loss in numbers and unique species is predicted to be in cryptofaunal organisms that associate with complex, hard substrate not found in soft sediments.

In the Gulf of California, commercial fisherman dive from small boats using hookah to harvest scallops and other benthic species. Observations after such fishing in rhodolith beds in Bahía Concepción in May 1991 revealed the almost complete extraction of adult scallops and extensive damage to the rhodolith beds caused by boat anchors, dragging of air hoses across the bottom, and direct diver disturbance while harvesting. Large areas of broken rhodolith fragments and areas cleared of living rhodoliths were common in previously intact beds (100 percent rhodolith cover) where fishing operations had taken place. The effects of trawling for shrimp in Gulf of California rhodolith beds have been photographed (Hall 1992), but the extent and severity are unknown. It is clear that trawling for bivalves in rhodolith beds reduces epifauna and infaunal diversity through large-scale homogenizing of the benthos (Hall-Spencer 1999).

It is not just the direct removal of thalli, but also the alteration of their morphology and size distribution that can affect species diversity and abundance. Thus, even disturbances with relatively low impacts may have severe effects at the community level. While direct impact of species harvest and the loss of habitat structure on species abundance can be measured, indirect effects are more difficult to estimate. For example, while the direct effects of trawling or bottom fishing for scallops and other benthic species decreases target species abundance, they also indirectly influence community structure by decreasing fragile or omnivorous species, increasing soft-sediment suspension-feeding species (Grall and Glemarec 1997), and increasing the number of scavengers (Hall-Spencer and Moore 2000). Long-term effects are difficult to estimate, but it is clear from extremely low coralline growth rates both worldwide and in the Gulf of California that recovery of the substrate after disturbance is likely to be very slow. Studies in the Gulf of California demonstrate that rhodolith beds are important in maintaining benthic species diversity and abundance, including economically important species, and the maintenance of high cover of intact, branching rhodolith thalli is essential to maintaining high diversity (Grall and Glemarec 1997; Grall et al. 2006).

As a consequence of our long-term experience working with rhodoliths, we have been able to identify at least eight species that occur primarily in rhodolith communities under the Mexican Official Norm to protect species (SEMARNAT 2001) listed in Riosmena-Rodríguez et al. (in press). Rhodolith beds are also habitats of special consideration because at least another 12 exploited species are usually found in this environment (Diario Oficial de la Federación 2004; Riosmena-Rodríguez et al., in press). In several parts of the world rhodolith beds (maerl) and rhodolith-forming species are listed as important to conserve as Special Areas of Conservation (United Kingdom and Europe) or are considered in the development of marine parks (BIOMAERL 1998). At present, stronger protection is difficult because there is no Official Mexican Norm for rhodolith beds. However, their preservation is becoming an essential part of marine conservation strategies in Mexico. Rhodolith beds are being considered in marine surveys. For example, in the original general model for designing marine reserves in the Gulf of California, Sala et al. (2002) used rhodolith beds as one of the relevant habitats for conservation. A few Marine Protected Areas (e.g., Loreto, Isla Espíritu Santo, San Pedro Mártir, and Revillagigedo) have included rhodoliths as habitats for conservation. Rhodoliths are increasingly being considered as part of the regional conservation planning of marine habitats in the Gulf of California (Ulloa et al. 2006; SEMARNAT 2007).

7 Contribution of Rhodoliths to the Generation of Pliocene-Pleistocene Limestone in the Gulf of California

Markes E. Johnson, David H. Backus, and Rafael Riosmena-Rodríguez

Geologists have been slow to understand the implications of crustose coralline algae as major contributors to the fabric of the rock record. To promote more research in this area, Adey and Macintyre (1973) summarized information on the distribution and ecology of modern coralline algae, demonstrated the range of key genera in geological time back to the Jurassic, and clarified the complex reproductive-cell terminology needed to identify both living and fossil species. They also devised a paradigm to show how changes in water depth, wave energy, and proximity to shore regulate the composition and volume of buildups by different coralline algae (fig. 7.1A). Accumulations covered by their model range from plated growth on rocky shores (or shallow ledges), encrustations nucleated around larger cobbles or smaller pebbles, and deposits derived from free-living (unattached) rhodoliths.

Our interest is limited to the Pliocene-Pleistocene record of fossil rhodoliths (with or without rock cores) and associated deposits characterized as rhodolith debris that accrued in the Gulf of California over the last 5.5 Ma. However, it is important to place this regional story in a wider geographic context and a longer span of geological time. Data on global diversification of coralline algae compiled by Aguirre et al. (2000) show that species diversity expanded gradually from the Late Jurassic (130 Ma) until the Early Miocene (about 20 Ma). Species diversity declined through the later Miocene and ensuing Pliocene epochs, but recovered markedly during the subsequent Pleistocene. Analysis by Halfar and Mutti (2005) emphasized how the rise in diversity of coralline algae to its Miocene peak came at the expense of coral-reef species that underwent a concomitant decline. They argued that rising nutrient levels associated with growth of the East Antarctic Ice Sheet boosted production of limestone enriched by rhodoliths. Miocene expansion of the ice sheet steepened the pole-to-equator temperature gradient in the Southern Hemisphere with attendant consequences toward intensification of oceanic circulation and more vigorous upwelling. The role of fleshy algae in competition with corals under rising nutrient levels is another factor that may have been significant, although fleshy algae have a very poor fossil record.

No explanation for the global Pliocene decline in coralline algae is obvious, other than onset of glaciation in the Northern Hemisphere at about 2.4 Ma, which may have caused environmental stress related to sea-level changes and climate-belt compression (Aguirre et al. 2000). The global diversity of coral species also continued to decline through the Pliocene (Halfar and Mutti 2005). Linkage between glaciation and declining rhodolith diversity is problematic, however, because similar factors related to expansion of the Antarctic Ice Sheet during the Miocene also favored diversification of coralline and fleshy algae. The Gulf of California is a semi-enclosed basin in a subtropical setting, where upwelling is generated by a combination of thermohaline circulation, tidal mixing, and zonal wind stress (Alvarez-Borrego and Lara-Lara 1991; Bray and Robles 1991). The massive rhodolith-rich deposits on Isla Carmen described by Eros et al. (2006), for example, suggest the central gulf area maintained levels of nutrient flow independent from climate changes related to ice advances in the Southern or Northern Hemispheres.

Previous work summarized by Foster et al. (1997) drew fundamental comparisons between the ecology of living rhodoliths and fossil counterparts in the Gulf of California. Building on this foundation, our goal is to consider three questions regarding rhodolith development framed by the gulf's Pliocene-Pleistocene record. To what extent are fossil rhodoliths preserved in limestone beds that show primary affinities to living rhodolith banks? How does limestone derived as a secondary product from rhodolith debris reflect the productivity of the original rhodolith populations? Finally, what can be said about the identification of

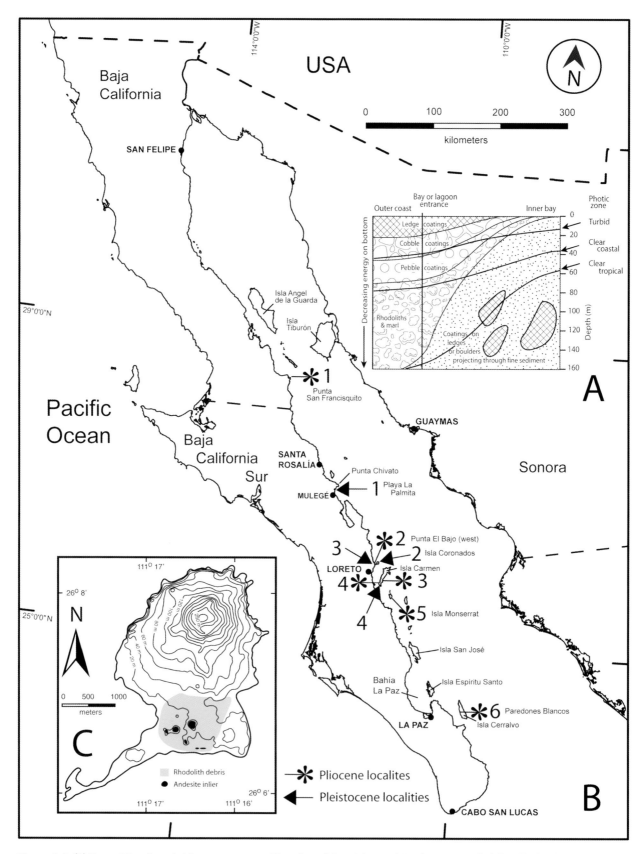

Figure 7.1. (A) Depositional model for crustose coralline algae (after Adey and Macintyre 1973); (B) Gulf of California map showing numbered localities for all Pliocene and Pleistocene deposits discussed in this chapter with abundant whole fossil rhodoliths or the debris derived from crushed rhodoliths; and (C) detailed inset map for Isla Coronados.

fossil rhodoliths to genus or species levels for the Gulf of California? Answers to the first and second questions also shed light on the local competition between reef corals and coralline algae recognized on a more global scale by Halfar and Mutti (2005). As cautioned by Adey and Macintyre (1973), productivity studies depend on reliable estimates of growth rates for different species. Fortunately, data of this kind are becoming available for the Gulf of California (Rivera et al. 2004; Steller et al. 2007a).

There are no published species descriptions for the abundant fossil rhodoliths from Pleistocene strata in Baja California studied by Dorsey (1997), Foster et al. (1997), or Cintra-Buenrostro et al. (2002) and from Pliocene strata studied by Eros et al. (2006). A comprehensive monograph covering these materials is needed to determine species composition using all known representatives in the fossil record. Such a treatment is beyond the scope of this chapter but would better facilitate stratigraphic analyses and the interpretation of distributional changes through the last several million years.

Appraising Past Rhodolith Contributions to Limestone Formation

The same LANDSAT and Advanced Spaceborne Thermal Emission and Reflection Radiometer (ASTER) satellite images that proved to be useful in tracking former shorelines around the Gulf of California (chap. 2) were employed in the systematic search for rhodolith limestone. For ETM+ data, bands 7,4,1 were combined to create composite images that showcase limestone deposits. Bands 8,3,1 proved to be the best three-band combination when using ASTER data. Although it is difficult to distinguish shelly marine limestone (chap. 9) from dune rock (chap. 11) using these images alone, the uniform composition of rhodolith limestone has a high reflectivity that typically shows up as bright beige patches.

Because the chief goal of this study is to appraise past rhodolith productivity in the formation of limestone deposits, it was critical to visit locations where total bedding thickness could be measured to help calculate basin volume. As part of this procedure, estimates on the degree of rhodolith concentration also needed to be made within a stratigraphic context. Sewell et al. (2007) developed the methodology to determine the population size of living rhodoliths required to produce a given deposit with a particular

level of purity. Until now, the only application of this method to analyze rhodolith limestone was undertaken by Ledesma-Vázquez et al. (2007a) with reference to deposits on Isla Coronados. The computations offered here significantly expand on the methodology of Sewell et al. (2007) to rate productivity in Pliocene-Pleistocene basins at several localities around the Gulf of California where rhodoliths are especially plentiful.

Whole fossil rhodoliths collected for identification were studied using petrographic thin sections and from observations using scanning-electron-microscope facilities at the Universidad Autónoma de Baja California Sur (UABCS). Taxonomic concepts and features are based on studies by Riosmena-Rodríguez et al. (in press). The fossil collections from this project are held at the Natural History Museum of UABCS.

Geographic Range of Gulf Rhodolith Deposits

Four Pleistocene deposits that date from the last interglacial epoch less than 125 Ka and six Pliocene deposits from about 5 to 2 Ma in age are covered in this survey. Generally, the Pliocene deposits track a broader latitudinal range in distribution than the Pleistocene deposits. The most northern latitude with abundant Pliocene rhodoliths presently known in the Gulf of California is from the Punta San Francisquito area at 28°26'N, while the southernmost latitude is from Paredones Blancos on Isla Cerralvo at 24°30'N (fig. 7.1B). The most northern latitude with abundant Pleistocene rhodoliths is from Playa La Palmita in the Punta Chivato area at 27°04'N, while the most southern latitude is from the southern end of Isla Carmen at 25°80'N (fig. 7.1B). Pleistocene deposits with a mixture that includes some rhodolith debris occur much farther south, for example, on Isla Espíritu Santo (Halfar et al. 2001), but not as a relatively pure facies.

Pleistocene Rhodolith Deposits
Playa La Palmita

Libbey and Johnson (1997) and Cintro-Buenrostro et al. (2002) described Upper Pleistocene deposits within a 1.5-km² area behind Playa La Palmita on Bahía Santa Inés immediately south of the Punta Chivato promontory (fig. 7.1B). Extending for nearly 2 km, low bluffs behind the beach have a maximum thickness of about 4 m and expose a poorly consolidated limestone rich in mollusk shells and rhodoliths. Many fossil rhodoliths from Playa La Palmita show fruticose morphology, but no detailed taxonomic descriptions have been made. The bluffs are dissected by small arroyos that can be followed 800 m inland to reveal the overall configuration of a shallow basin. Where the deposit wedges out at the rear of the basin, andesite cobbles with heavy encrustations of platy coralline algae mark the former shoreline. Dense shell accumulations with occasional whole rhodoliths form the inner part of the basin, which is floored by Pliocene siltstone now exhumed in some of the arroyo bottoms. The southern margin of the basin is marked by an abrupt abutment unconformity, where an adesite ridge borders extensive rhodolith debris (Libbey and Johnson 1997, fig. 5A). Johnson (2002a, fig. 4) illustrated the abundance of fragmented rhodoliths from this zone, in which whole rhodoliths are rare and shelly material is almost entirely absent. This zone represents the outer, seaward part of the basin and is estimated to cover an area of approximately 0.75 km² with an average thickness of 3 m (for 2.25 million m³).

Following the experimental work of Sewell et al. (2007), it takes 16,265 whole rhodoliths with an average diameter of 5 cm to make 1 m³ of debris composed of sand-size fragments. By this accounting, it would require more than 36 billion crushed rhodoliths to produce the relatively pure rhodolith debris at the core of the Playa La Palmita basin. Considering that the largest whole rhodoliths recorded from this site are only 4 cm in diameter (Johnson 2001, fig. 4), the actual population number would be much greater. Furthermore, rhodolith populations living in the Gulf of California tend to be dominated by small individuals (Steller et al. 2003), making realistic estimation of population numbers more difficult. Since 1995, commercial development has resulted in the construction of many seaside cottages on Playa La Palmita and the best Pleistocene exposures are no longer accessible.

Isla Coronados

The same rough calculations can be made with respect to extensive Upper Pleistocene deposits of crushed rhodolith debris on Isla Coronados, located 11 km

Figure 7.2. Looking north with Pleistocene volcanic cone of Isla Coronados in the background and a thick bioclastic deposit composed of rhodolith debris in the foreground. The beds accumulated as over-wash deposits within a protected lagoon about 121 Ka, during the last interglacial epoch. Photo credit: Markes E. Johnson

northeast of Loreto (fig. 7.1B, C). Isla Coronados is part of the Parque Nacional Bahía de Loreto, which means that environmental preservation is assured. Johnson et al. (2007) mapped the confluence of two shallow canyons on the southern side of Isla Coronados, where a 121-Ka coral reef is well exposed in the sides of the canyon walls. The exposed margins of the reef outline an area covering 1.6 ha (Johnson et al. 2007, fig. 3). Many of the corals can be shown to have colonized a hard substrate consisting of andesite boulders and cobbles that paved a great lagoon on the leeward side of the Pleistocene volcano dominating Isla Coronados. Below the pavement is a 6-m-thick accumulation of crushed rhodolith debris organized as individual beds from 40 to 60 cm in thickness (fig. 7.2). Ledesma-Vázquez et al. (2007a) studied these beds, which show a uniform 20° dip to the north toward the Pleistocene volcano, and concluded they originated as over-wash deposits related to episodes of storm surge that transported broken rhodolith material from outside the lagoon across a barrier of small andesite islands now exhumed by erosion. Based on Sewell et al. (2007), it may be calculated that the highly pure deposit of rhodolith hash within the 1.6 ha would require an original population of more than 1.5 billion rhodoliths with an average diameter of 5 cm to accrue foreset beds 6 m thick. The larger area with rhodolith debris that surrounds the canyon on Isla Coronados has a surface area of 50 ha (fig. 7.1C). If rhodolith debris within this area is 6 m thick, then more than 48 billion whole rhodoliths would have been needed to make the full lagoonal deposit (Ledesma-Vázquez 2007a).

Whole rhodoliths are extremely rare inside the lagoon complex. On the outer seaward side of the andesite barrier, however, whole rhodoliths are commonly found in Pleistocene limestone beds that include a diverse mollusk fauna. Nearly half the rhodoliths sampled from this area exhibit foliose (bladed spherical morphology) to fruticose thalli (i.e., delicately branching spherical morphology) and preserve secondary pits between filaments, flat or rounded epithelial cells, and conceptacles with uniporate tetrasporangia diagnostic of species belonging to *Lithophyllum*. A quarter of the sample also shows fruticose morphology but is differentiated by cell fusions between filaments, flared epithelial cells, and multiporate conceptacles diagnostic of *Lithothamnion*. Still other specimens with fruticose morphology exhibit cell fusions and secondary pits between filaments, as well as flared epithelial cells and sori typical for

species of *Sporolithon*. The largest rhodolith in the collection has a diameter of 6.5 cm.

Punta El Bajo

The best example of a fossil deposit comparable to a living rhodolith bank is found in Pleistocene cliffs that reach 3 m in height near Punta El Bajo on the peninsular mainland west of Isla Coronados (fig. 7.1B). Coincidently, the east-facing intertidal zone below the same sea cliffs is a place where numerous modern rhodoliths identified as *Lithothamnion muelleri* are recorded to have washed ashore (Sewell 2007, fig. 2). A distinct color change from more beige and finely textured limestone in the lower part of the outcrop to darker and more coarsely textured limestone in the upper layers is evident from the locality photo (fig. 7.3A). The entire outcrop is dominated by fossil rhodoliths and rhodolith debris. Fossil shells are notably absent. Whole rhodoliths typically with a fruticose morphology are common in the lower part of the outcrop, but they are scattered through a bioclastic matrix dominated by crushed rhodolith debris. At this level, the limestone is massive and shows no signs of bedding, which implies that the deposit underwent reworking by waves or currents or suffered bioturbation by infaunal members no longer evident.

Above the boundary marking the color change sits a 20-cm thick layer of whole rhodoliths, densely packed and dominated 2:1 by foliose thalli over lumpy thalli (fig. 7.3B). The many bladed rhodoliths can be assigned to the extant species *Lithothamnion margaritae*, but some lumpy samples from this bed also exhibit features characteristic of species belonging to *Sporolithon*. On average, thallus diameter is 4 cm and the density of whole rhodoliths is 7 to 8 thalli per 10 cm^2 as viewed in stratigraphic cross section. This particular bed is continuous over a lateral distance of 30 m along the sea cliff. Approximately 4,500 rhodoliths are exposed in this layer. Clearly, the cliff face is being cut back by wave erosion and may have retreated as much as 10 m based on the width of the intertidal shelf extending seaward from its base. Without core samples, there is no way of knowing how far inland the layer extends. Above the layer with densely aggregated rhodoliths, the color remains the same but rhodoliths become more diffused through the limestone. Foster et al. (1997, p. 136) mentioned a profile from the same vicinity and suggested that the scenario is "very similar to the vertical stratification observed in the middle of living wave beds." However, the dark

Figure 7.3. (A) Looking west on a Pleistocene outcrop exposed at the shore near Punta El Bajo 11 km north of Loreto, with Jorge Ledesma-Vázquez for scale; and (B) close-up view showing details of a 20-cm-thick layer bearing abundant, whole rhodoliths with an average diameter of 4 cm. Photo credits: David H. Backus

anoxic sediments on which rhodoliths in living wave beds typically are found in Bahía Concepción are not reflected in the Punta Bajo Pleistocene profile. A comparable pattern of whole rhodoliths stratified on fragmented rhodoliths also occurs in modern settings related to deeper current beds in which there is no trace of fine siliciclastic sediment or signs of anoxia (M. Foster, personal communication 2007). Thus, the fossil scenario near Punta Bajo may be more characteristic of accumulation in a deeper-water current bed than a shallow-water wave bed.

As reviewed by Johnson et al. (2007, p. 116), the sea cliff shown in figure 7.3 can be followed north for about 1 km to Punta Bajo, where it is unconformably overlain by a Pleistocene coral reef tentatively dated to an age of 105 Ka (isotope substage 5c) or possibly 80 Ka (isotope substage 5a).

Southern End of Isla Carmen

Low carbonate ramps occupy nearly 3 km² on the southern end of Isla Carmen (fig. 7.1B). Here, limestone beds with a minimum exposed thickness of 17 m are composed of 75 to 95 percent rhodolith debris. Bivalves that make up a small fraction of the deposits include *Codakia distinguenda* and *Periglypta*

multicostata, which occur articulated and planted in growth position within the rhodolith debris. No rhodolith samples from this area were available for detailed taxonomic study. As shown on the satellite image for southern Isla Carmen (fig. 7.4), two beige patches are partitioned along a north–south axis by a thin, darker strip representing a former island. Unlike Isla Coronados, the configuration of these Pleistocene ramps on Isla Carmen suggests that rhodolith debris did not accumulate in a closed lagoon but was transported into a protected setting under shoal-water conditions. Approximately 50 million m³ of rhodolith limestone with an average purity of 85 percent are in place on the southwest- and southeast-oriented ramps of Isla Carmen. Following the formulation by Sewell et al. (2007) based on rhodoliths with a diameter of 5 cm, it would require a population of 700 billion rhodoliths to contribute skeletal material to these ramps.

At least two terraces were cut into the rhodolith ramps of southern Isla Carmen. The terraces can be followed laterally at elevations of 6 and 11 m above present sea level, respectively. In places, dense congregations of *Porites* corals grew in situ on the terraces (Johnson et al. 2007). Assuming that the 11-m terrace is the usual regional marker for the last interglacial epoch (isotope substage 5e), then the rhodolith ramps must be older than 125 Ka.

Figure 7.4. False-color satellite image for the southern part of Isla Carmen in the lower Gulf of California showing limestone deposits dominated by rhodolith debris in three regions outlined in red: Arroyo Blanco basin on the eastern coast, Bahía Marquer basin on the western coast, and the southern end. Image acquired from the Pan-American Center for Earth and Environmental Studies. Original data set was collected by the NASA LANDSAT Program, LANDSAT TM scene LT40350420089161, L1G. USGS, Sioux Falls, June 10, 1989. This image was resampled (Gram-Schmidt spectral sharpening) using band 8 from a second data set collected by NASA LANDSAT Program, LANDSAT ETM+ scene L7CPF20011001_20011231_05, L1G, USGS, Sioux Falls, October 17, 2001, and acquired from the University of Maryland Global Land Coverage Facility.

Pliocene Rhodolith Deposits
Punta San Francisquito

Briefly discussed in chapter 3, the narrow channel that enters the Pliocene basin at San Francisquito (fig. 7.1B) features a north-dipping ramp on granodiorite with a 1.5-m-thick basal conglomerate derived from eroded granodiorite clasts. This is the only known locality in the Gulf of California where Pliocene rhodoliths are encrusted on pebbles. The rhodoliths measure 5 cm across with calcareous rinds that completely enclose granodiorite pebbles up to 3 cm in diameter. Whole rhodoliths float in a coarse matrix consisting of silica and plagioclase granules mixed with abraded rhodolith debris. Based on surveys using a 50-×-50-cm grid set up in stratigraphic profile on the upper part of the conglomerate, 25 rhodoliths typically fit into a 2,500-cm^2 cross section (1 per 10 cm^2). The lateral continuity of the basal conglomerate cannot be appraised due to outcrop limitations, which means that population size is difficult to estimate. Ecologically, however, the setting clearly invokes an exposed rocky shore in which wave energy kept pebbles in almost continuous motion. Unfortunately, no rhodolith samples from this site were available for detailed taxonomic study. The only other example of this kind from Baja California comes from a Cretaceous island on the Pacific coast, where rhodoliths belonging to the genus *Sporolithon* encrust andesite pebbles (Johnson and Hayes 1993).

West of Punta El Bajo

Reviewed in detail in chapter 3 for the insight it provides on coastal tectonics in the Loreto region, a sequence of Upper Pliocene strata with repetitive couplets of andesite conglomerate and sheeted calcarenite composed of rhodolith debris is exposed west of Punta El Bajo (fig. 7.1B). Four conglomerate beds are overlain in succession by four calcarenite beds that range from 5 to 10 m in thickness (Dorsey 1997). In each couplet, the conglomerate represents the nearshore deposit eroded seaward from a rocky coastline and the calcarenite signifies an offshore deposit that shifted landward with debris derived from nearby rhodolith banks. The transition between successive couplets is remarkably sharp and indicates local cycles of rapid coastal uplift during the Late Pliocene. The ramp configuration of each couplet showing increasing steepness is very clear as exposed in longitudinal profile in the sea cliffs west of Punta El Bajo, but this arrangement gives no hint as to the lateral extent of each calcarenite unit. Therefore, it is not possible to estimate the population numbers for whole rhodoliths from which the deposits were derived. No rhodolith samples from this area were available for detailed taxonomic study.

Arroyo Blanco on Isla Carmen

The largest concentration of rhodolith debris yet discovered in the Gulf of California is also the deposit with the longest history of continuous development, spanning much of the Pliocene into the Early Pleistocene (Eros et al. 2006). The site is the Arroyo Blanco basin on the eastern side of Isla Carmen (fig. 7.1B). Based on our re-evaluation of the basin perimeter using the satellite image in figure 7.4, the surface area covered by the Arroyo Blanco basin is 3.87 km^2. As measured in the canyons of Arroyo Blanco, the exposed thickness of the basin's ramped strata amounts to 157 m. The base of the sequence is not exposed in the deepest part of the basin at the mouth of Arroyo Blanco, thus maximum thickness is unknown. Based on a bed-by-bed analysis of fossil content, the exposed basin sequence was evaluated as 64 percent filled by calcarudite and calcarenite derived from crushed rhodoliths (Eros et al. 2006). Whole rhodoliths are rare, do not exceed 4 cm in diameter, and demonstrate a fruticose morphology. Photographs from Eros et al. (2006, figs. 3A, B) illustrate a typically massive, 9.6-m-thick unit highly enriched in rhodolith debris from the lower part of the succession showing both lateral perspective from afar and the coarse texture of the accumulation close-up.

Crudely trapezoidal in outline, the shorter sidewalls of the Arroyo Blanco basin are marked by steep normal faults and the long rear margin of the basin, where sedimentary deposits pinch out against Miocene andesite, is regarded as an unconformity-based rocky shoreline. Derivation of basin volume is complicated, but it can be assumed that the basin's simple ramp construction is footed on a sloping contact with underlying Miocene volcanics. Thus, the basic shape of the basin is prismatic with a wedge-shaped cross section. Multiplication of surface area by one-half the minimum thickness of the basin sediments provides the volume of the prism: 303,795,000 m^3. Only a portion of the basin sediments (64 percent) is derived from rhodolith debris, which amounts to 194,428,800 m^3. Following the formulation by Sewell et al. (2007)

Figure 7.5. Looking south across Bahía Marquer on the western coast of Isla Carmen (see figs. 7.1B and 7.4 for location). Sea cliffs in the foreground are approximately 17 m high. The rose-tinted beds exposed near sea level are densely packed with rhodolith debris in which whole rhodoliths are rare. Photo credit: Patty Liao

based on rhodoliths with a diameter of 5 cm, it would necessitate a population of more than 3 trillion rhodoliths to contribute skeletal material to the Arroyo Blanco basin spread over approximately 5 Ma. The actual number should be much higher, due to the fact that the largest whole rhodolith recovered from the basin is only 4 cm in diameter and typical populations are dominated by individuals in the 1- to 2-cm-diameter ranges. No rhodolith samples from this area were available for detailed taxonomic study.

Bahía Marquer on Isla Carmen

The western side of Isla Carmen features another large basin (3.42 km²) that provides a symmetric counterbalance to the Arroyo Blanco basin on the eastern side (figs. 7.1B, 7.4). A view to the south across the front of the Bahía Marquer basin shows Upper Pliocene stratification exposed in sea cliffs (fig. 7.5). Like the Arroyo Blanco basin, the full thickness of strata in the Bahía

Marquer basin is unknown because basement rocks are not exposed in arroyos that cut through the thickest part of the succession. Strata exposed in the sea cliffs vary in thickness from 17 m in the north to 20 m in the south. Our field studies indicate that the lower 10 m of the succession (visible as a rose-colored limestone in fig. 7.5) is enriched in rhodolith debris up to 60 percent by volume. Whole rhodoliths with a fruticose morphology are only locally common in these beds, but the maximum diameter does not exceed 4 cm. Half the succession in sea cliffs at the southern end of the bay (amounting to a thickness of 11.5 m) is formed by massive limestone composed of calcarenite, 90 percent of which is derived from rhodolith debris. There are no whole rhodoliths preserved in this unit. The massive rhodolith limestone in this part of the bay is capped by reef limestone 3.25 m in thickness and thought by Durham (1947, p. 11) to be Pliocene in age. In our exploration of the basin, *Porites* corals were found extensively throughout the center of the basin at an elevation about 70 m above present sea level. The age of this widespread coral capstone remains to be tested by radiometric dating. What is certain is that approximately 10 m of rhodolith-rich limestone of undisputed Pliocene age has a wide distribution in the Bahía Marquer basin.

The calculations are more tenuous, but assuming that a 10-m-thick rhodolith bed has an average purity of 75 percent and is evenly distributed throughout the basin, then it would be the equivalent of 25,650,000 m³ of pure rhodolith debris. Again following the formulation of Sewell et al. (2007) using 5-cm-diameter rhodoliths, it would take the contributions of more than 400 billion whole rhodoliths to generate the massive layer. As before, the actual number would be much higher due to the fact that smaller rhodoliths no more than 4 cm in diameter are known from this unit and typical living populations are dominated by smaller individuals (Steller et al. 2003). No rhodolith samples from this area were available for detailed taxonomic study.

Isla Monserrat

Situated 45 km southeast of Loreto (fig. 7.1B), Isla Monserrat is a small island (18.5 km²) with a Pliocene history as an even smaller paleoisland surrounded by a shallow marine shelf (see chap. 3). The geological map compiled for this volume (fig. 7.6) suggests the island underwent two distinct episodes of faulting with an earlier phase dominated by north–south normal faults that offset Middle Pliocene limestone at about 2.7 Ma and a subsequent phase of northwest–southeast faulting congruent with the rift-and-transform system currently active in the gulf. Reactivation of the north–south faults due to this later phase of transtensional faulting facilitated the island's extraordinary tectonic uplift. As a result of faulting, the Middle Pliocene carbonate shelf that formerly covered much of Isla Monserrat was sliced, shifted vertically, and reduced by erosion to a surface area equal to only 9 percent of the present island.

With regard to Pliocene rhodoliths, the most striking feature on Isla Monserrat is a rimmed plateau formed by a veneer of limestone no more than 3 m thick covering 35.6 ha in the north-central part of the island (fig. 7.6). The limestone plateau sits as an angular unconformity on Miocene andesite with layered flows that dip 35° east (fig. 7.7A). The plateau's rim varies in elevation from 185 to 193 m above present sea

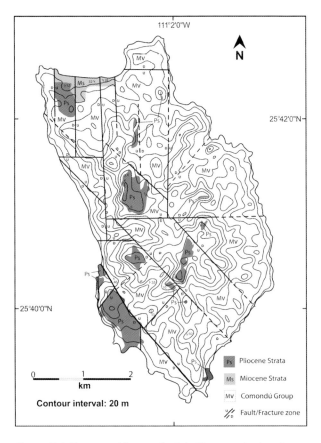

Figure 7.6. Topographic map for Isla Monserrat, showing principal faults and the distribution of Pliocene limestone with distinct facies that separate rhodolith deposits from shell deposits (see fig. 7.1B for location).

level, and the limestone layers exposed on the circumference dip from between 6° and 8° uniformity toward the center from all sides (fig. 7.6). In effect, the plateau preserves a shallow basin elevated by postdepositional faulting around its margins to a considerable height commanding the northern half of the island.

Many pectens (*Lyropecten subnodosus* and *Argopecten abietis*) are found around the rim of the plateau in a limestone matrix dominated by the debris of crushed rhodoliths. Whole rhodoliths are absent from the rim rocks but commonly occur at the center of the plateau, where they have been eroded from the host rocks. Whole rhodoliths collected from the plateau are illustrated in figure 7.7B, showing a selection of morphologies including lumpy, foliose, and fruticose thalli. Lumpy rhodoliths (upper row) are the most robust members of this facies and attained a maximum diameter of 3.85 cm. These show detailed characteristics such as flared epithelial cells and sori that indicate *Sporolithon* species. Foliose rhodoliths are rare

Figure 7.7. (A) Looking north across a 194-m-high plateau on Isla Monserrat showing rhodolith limestone of Pliocene age (beige) resting unconformably on andesite (red) of Miocene age; and (B) the range in morphology of whole fossil rhodoliths collected from the plateau. Photo credits: Markes E. Johnson

(second row), but attained a maximum diameter of 3.73 cm. These can be attributed to species of *Litho-phyllum*. Fruticose rhodoliths are the most abundant (bottom two rows); they show a maximum diameter of more than 4 cm but an average diameter of about 2.5 cm. Some of these exhibit detailed characteristics of *Lithothamnion* species. Ecologically, the setting represents a shallow depression in the Pliocene sea-bed where rhodoliths proliferated under the stimulation of gentle to moderate wave agitation.

Isla Cerralvo

The most southern of the major Pliocene rhodo-lith deposits in the Gulf of California is located on the western side of Isla Cerralvo at Paredones Blancos (fig. 7.1B). Hertlein (1966, p. 272) photographed a coastal outcrop at this locality showing a "white stratum composed of calcareous algae of Pliocene

age." Our investigation of this locality confirms that the stratum is a massive deposit with a concentration of rhodolith debris in excess of 80 percent in purity. The unit is 10 m in thickness. Likewise, the original description by Hertlein (1966) gives no sense of the outcrop's lateral continuity along the shoreline, but we traced it for a distance of about 0.75 km. Due to the faulted relationship of the deposit to more inland rocks, it is impossible to determine its volume. The significance of the locality is most clearly related to its occurrence in the lower Gulf of California as a former embayment where rhodolith debris became trapped.

Summary of Major Relationships

Among the ten fossil deposits described in this review, only three involve large numbers of whole rhodoliths buried and preserved under conditions that resemble those in which crustose coralline algae live today. The

Pliocene site near Punta San Francisquito reflects a high-energy rocky shore in which rhodoliths fully encrusted granodiorite pebbles eroded from nearby sea cliffs. Abundant whole rhodoliths from Isla Monserrat occupied a shallow Pliocene basin on an elevated fault block off the peninsular mainland of Baja California. South of Punta El Bajo, a sea cliff of Pliocene age preserves a stratum of densely clustered rhodoliths that appears little different than the concentration of rhodoliths that live in the adjacent, current-swept channel between the peninsular mainland and Isla Coronados today. In terms of the model developed by Adey and Macintyre (1973, fig. 31) to illustrate how changes in water depth, wave energy, and proximity to shore regulate the spatial deployment of rhodoliths, all three fossil deposits can be pegged to a narrow range of near-shore, shallow-water conditions.

In contrast, all other Pleistocene and Pliocene deposits described here represent transported debris derived from thoroughly crushed rhodoliths. This category is not explicitly covered in the model of Adey and Macintyre (1973) reproduced in figure 7.1A, although the marl component in their model could be construed as part of normal rhodolith disintegration. In the Gulf of California, fossil debris deposits tend to be trapped in bays along north–south coasts (such as the Arroyo Blanco and Bahía Marquer embayments on Isla Carmen) or on the southern side of islands or promontories (such as Punta Chivato and Isla Coronados). The tombolo and associated lagoons at El Requesón in Bahía Concepción represent a contemporary repository of concentrated rhodolith debris (Hayes et al. 1993). So long as the sand-size detritus of crushed rhodoliths remains unstabilized, it is difficult for corals to colonize the substrate. Once rhodolith debris becomes cemented or a hardground consisting of a cobblestone pavement is laid on top of rhodolith debris, *Porites* growth is abetted in the same sheltered settings (Johnson et al. 2007).

In terms of volume, the amount of rhodolith debris incorporated as Pliocene or Pleistocene limestone can be enormous, implying that addition of fresh material was continual. The rarity of corals in many basins suggests that stabilization due to burial by the influx of new material generally preceded lithification at depth. By far the greatest volume of fossil rhodolith debris found anywhere in the Gulf of California is from Arroyo Blanco on Isla Carmen. If the 3 trillion rhodoliths estimated to have contributed detritus to the basin were prorated on an annual basis over the roughly 5-Ma lifetime of the basin, the yearly input amounts to a paltry 600,000 whole large rhodoliths. Taking growth rates into account, a rhodolith 5 cm in diameter may take about 40 years to mature based on the field and laboratory work in the southern Gulf of California by Rivera et al. (2004). Making the assumption of sediment recruitment from rhodoliths on a 40-year storm cycle, however, does not change the picture significantly. We conclude that the potential for sediment production from rhodoliths was stupendous by any measure during any interval of time through the last 5.5 Ma in the Gulf of California. The limiting factor in the generation of rhodolith limestone was not controlled by population size but by the volume of potential sedimentary traps in the form of available lagoons or protected embayments. In addition to sequestration of rhodolith material as Pliocene-Pleistocene limestone, it can be assumed that large volumes of detritus were continually swept offshore into deeper gulf waters where dissolution occurred.

8 Ecological Changes on the Colorado River Delta
The Shelly Fauna Evidence

Guillermo E. Avila-Serrano, Miguel A. Téllez-Duarte, and Karl W. Flessa

The Colorado River delta is a very important breeding site for valuable fisheries that include species such as the white shrimp (*Litopenaeus stylirostris*), brown shrimp (*Farfantepenaeus californiensis*; Galindo-Bect 2003), and gulf corvine (*Cynoscion othonopterus*; Rowell et al. 2005). In addition, the delta is a restricted habitat for endangered species, such as the porpoise vaquita (*Phocoena sinus*), totoaba fish (*Totoaba macdonaldi*; Millán-Núñez et al. 1999; Pitt et al. 2000), and the bivalve *Mulinia coloradoensis* (Flessa and Téllez 2001; Rodríguez et al. 2001a; Cintra-Buenrostro et al. 2005; Avila-Serrano et al. 2006). In fact, the delta has become one of the world's most severely affected ecosystems. In less than 100 years of intensive human use and modification of the river's flow by dams and irrigation projects, the delta's marine biota has experienced structural changes, loss of biodiversity, ecosystem degradation attributed to the cessation of freshwater runoff to the Gulf of California (Carrillo et al. 2001), and a steady rise in salinity (Cohen and Henges-Jeck 2001).

Because no checklist of the delta fauna prior to the ecosystem collapse is available, many unknown species may have disappeared. Unfortunately, nektonic organisms are poorly preserved in intertidal environments due to taphonomic processes that frequently destroy skeletal remains. For that reason, the durable shelly intertidal fauna provides a reasonable basic source to investigate the estuary's former benthic ecology, when abundant fresh river water mixed with marine salty water from the Gulf of California. For example, Kowalewski et al. (2000) estimated that 2 trillion shells belonging to the once-abundant and widely distributed estuarine bivalve *M. coloradoensis* (fig. 8.1) were concentrated at the mouth of the Colorado River prior to the damming of the river. Actually, small surviving populations of this endemic species still live in restricted sites near the river mouth (Flessa

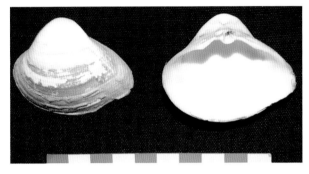

Figure 8.1. External and internal views of the bivalve *Mulinia coloradoensis* (scale = 10 cm). Photo credit: Miguel A. Téllez-Duarte

and Téllez 2001). These few populations are considered characteristic of the region's ecology prior to alteration of the estuary.

In this sense, recent and ancient shelly fauna deposits provide an opportunity to compare the past ecosystem with the living one and improve our understanding of the alteration process from an ecological and paleoecological approach. In addition, archeological shell middens left behind by native peoples may provide another source of information regarding early human impact on coastal resources and the environment (Téllez-Duarte et al. 2001a, 2001b). In this chapter we summarize our knowledge of comparatively recent ecological changes in the Colorado River delta, with a focus on the intertidal-mudflat ecosystem.

Geographic and Climatic Framework

The marine portion of the Colorado River delta is located within the Reserva de la Biósfera Alto Golfo de California y Delta del Río Colorado, situated from 31°01'to 32°22'N and from 112°59' to 115°13'W along the coastline between San Felipe, Baja California, and El Golfo de Santa Clara, Sonora (fig. 8.2).

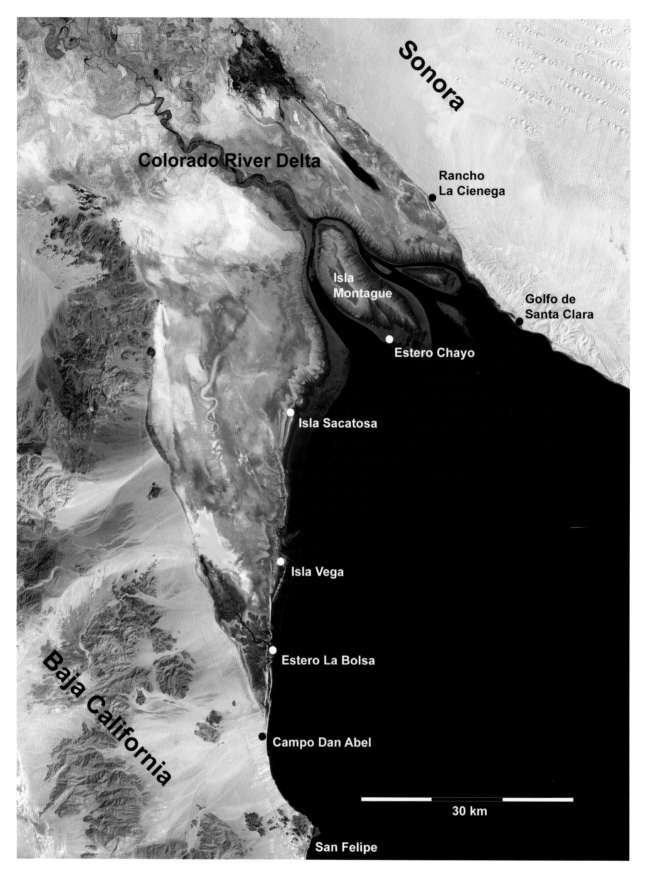

Figure 8.2. Satellite image showing the Colorado River delta and North Gulf Region with the sampling localities discussed in the text. White strings along the coast are cheniers separated by mudflats. Image acquired from the University of Maryland Global Land Coverage Facility. Original data set was collected by NASA LANDSAT Program, LANDSAT ETM+ scene L7CPF20000401_20000630_09, L1G, USGS, Sioux Falls, 04/10/2000.

The Colorado River drains an approximate area of 630,000 km² and runs along a course approximately 2,300 km in length from the Rocky Mountains in the United States to the mouth of the Gulf of California in Mexico (Carrillo et al. 2001).

The delta region is hot and arid, with an annual average rainfall of 68 mm, monthly mean air temperatures that range between 14 and 34°C (Miranda-Reyes et al. 1990), and surface tidal-flat temperatures between 3.3 and 40.6°C (Goodwin et al. 2001). A macrotidal regime with maximum tidal amplitudes of 12 m occurs at the river mouth (Thompson 1968). The very large tidal range and low regional slope result in tidal flats that extend seaward by more than 2 km at low tide. Sediments are typically composed of fine sand and silt in the south and silt and mud in the north (Thompson 1968). High evaporation rates cause increased salinity, where denser water sinks and flows to the south and is replenished by surface water flowing north (Lepley et al. 1975).

Geologic Framework

The structural setting of the Colorado River delta area is located in the southern part of the Basin and Range Province, which is bounded by the main fault zones in the Elsinore, San Jacinto, and San Andreas Fault systems. Here, down-faulted grabens have created natural basins where thick beds of sediments eroded from the adjacent uplands accumulate. These sedimentary units are locally interbedded with lava and tuffs from volcanic activity. The most conspicuous feature is the Salton Trough, which covers an area approximately 200 km long by 150 km wide. The thickness of sediment within the Salton Trough exceeds 6 km in the deepest part (Dibblee 1996). The geologic history in the area is structurally and sedimentologically complex, with extensive faulting after the Late Miocene rifting of the North American plate and evidence of an initial seawater incursion that occurred between 16 and 13 Ma as part of the formation of the proto-Gulf of California. Two other flooding events occurred during the Late Miocene to Early Pliocene, from 9 to 6 Ma and 5 to 3 Ma, whose deposits belong to the Imperial Formation (Carreño and Smith 2007). To the east, the Imperial Formation is correlative with the Late Miocene Bouse Formation, which defines the eastern limit of the Salton Trough (Damon et al. 1978; Shafiquallah et al. 1980).

First arriving from the eastern side of the Colorado River, sediments in the Salton Trough were deposited at the beginning of the Pliocene, about 4.3 Ma (Johnson et al. 1983). Probably the earliest river-derived sediments recorded are located on the eastern side of the basin in the Bouse Formation. Their age has not been dated directly, but it can be bracketed between a 5.5-Ma tuff from the base of the Bouse Formation and a 3.8-Ma basalt that overlies undated Colorado River gravels widely distributed from the Gulf of California to the Lake Mead area (Shafiquallah et al. 1980; Buising 1990).

Fossiliferous Pleistocene Colorado River gravels and sandstones tectonically uplifted by the San Jacinto–Cerro Prieto Fault system contain an abundant and diverse fauna of vertebrates and silicified wood (Shaw 1981). They are well exposed in the Golfo de Santa Clara of Sonora, Mexico. Unnamed deltaic mudstones of this age are exposed at Estero La Bolsa and Punta El Faro (fig. 8.3). Holocene sediments are predominately fine-grained clastics that consist of silts and fine sands deposited on progradational tidal flats, where longitudinally oriented biogenic coarse-grained beach bars (*cheniers*) are common. These deposits are almost completely dominated by mollusk shells concentrated by waves and tides, as described in detail by Thompson (1968). Facies development in the delta area has been controlled by the activity of the Colorado River. When the river was flowing, input of a large volume of fine-grained sediment accumulated and tidal flats prograded seaward. When the flow is low or null, as today, the fine sediment is reworked and eroded away. Under those conditions, the coarse fraction becomes concentrated along the shoreline as shell bars or cheniers composed almost entirely of the clam *M. coloradoensis* (Sykes 1937; Thompson 1968).

Figure 8.3. Pleistocene deposits around El Faro. Photo credit: Miguel A. Téllez-Duarte

Living Shelly Fauna and Shell Beds
Living Fauna

Little is known about the living shelly fauna native to the Colorado River delta area. In their study of the living and dead fauna on the intertidal tidal flats from the river's mouth to the lower delta limit near Campo Don Abel (fig. 8.2), Avila-Serrano et al. (2006) noted seasonal variations and reported a total of 26 species (table 8.1, data and sampling localities available at www.geo.arizona.edu/ceam/data1/). Mollusks are the dominant component of the intertidal fauna, with only one species of echinoid (*Mellita longifissa*) and one brachiopod (*Glottidia palmeri*), which are rare year-round. Besides being scarce and strongly dominated by a few species, the fauna is not uniformly distributed. Only three species make up 75 percent of all individuals in the whole area. Only one species, the epifaunal gastropod *Nassarius moestus*, is abundant throughout the year. Other relatively abundant species include the shallow-infaunal bivalve *Tellina meropsis* and the shallow-infaunal echinoid *M. longifissa*. The shallow-infaunal *M. longifissa* is abundant only in the southern part of the delta at Campo Don Abel, together with the shallow-infaunal bivalves *Donax navicula* and *Chione pulicardia* and the deep-infaunal bivalve *Tagelus affinis*. The infaunal brachiopod *G. palmeri* is characterized by a very patchy distribution from the vicinity of the mouth of the Colorado River (Avila-Serrano et al. 2006) south to Campo Don Abel (Kowalewski 1995). The once-dominant bivalve *M. coloradoensis* (fig. 8.1) is very rare, occurring only in the vicinity of the river mouth. Only one individual was reported at Campo Don Abel (Avila-Serrano et al. 2006), but the species is virtually extinct south of Isla Vega (fig. 8.2). Another distinctive species in the delta area is the bivalve *Chione cortezi*. This species has been confused with the widely distributed *Chione fluctifraga*, but it appears that *C. cortezi* is restricted to the delta area from Playa Paraiso north of San Felipe to the vicinity of the river mouth (Villarreal-Chávez et al. 1999).

Death Assemblages

Shell-bed pavements are common on the intertidal mudflats in the delta area (fig. 8.4). Nevertheless, no previous attempts have been made to describe their faunal composition. To remedy this situation, samples were collected by wet-sieving sediment through a 6-mm mesh at five stations located around the Colorado River delta. The stations are located from north to south at the Golfo de Santa Clara Biological Station on the Sonoran coast and at Estero Chayo, Isla Sacatosa, Isla Vega, and Campo Don Abel on the Baja California coast. At each station, transects were made perpendicular to the shoreline along the tidal flat during low tide. Surface samples comprised the uppermost 10 cm and an area of 1 m² was sampled. The shelly fauna was cleaned in the laboratory. All

Table 8.1. Checklist of living shelly fauna from the Colorado River delta.

Species	Total	%
Nassarius moestus	720	36.85
Tellina meropsis	430	22.01
Mellita longifissa	320	16.37
Chione fluctifraga	214	10.95
Glottidia palmeri	111	5.68
Donax navicula	45	2.30
Chione pulicardia	38	1.94
Mulinia coloradoensis	15	0.77
Encope grandis	11	0.56
Pitar concinnus	6	0.31
Strigilla interrupta	6	0.31
Natica chemnitzii	5	0.26
Solenosteira capitanea	4	0.21
Nuculana impar	4	0.21
Protothaca grata	4	0.21
Tagelus affinis	4	0.21
Tellina amianta	3	0.15
Cumingia pacifica	2	0.10
Epilucina californica	2	0.10
Felaniella sericata	2	0.10
Macoma siliqua siliqua	2	0.10
Unidentified bivalve sp. 1	1	0.05
Olivella zanoeta	1	0.05
Saxicava pacifica	1	0.05
Veneridae	1	0.05

Figure 8.4. Shell beds on mudflats of the Colorado River delta. Photo credit: Miguel A. Téllez-Duarte

complete and fragmented specimens were counted. Fragmented specimens were considered to be complete only if the umbo area of bivalves was intact or the tip of gastropods was retained. Specimens were identified according to Keen (1971) and Keen and Coan (1974). Those species more difficult to recognize were compared with reference materials in collections housed at the University of Arizona and the San Diego Natural History Museum.

In total 101 stations were sampled, with 779 individuals divided among 115 taxa (tables 8.1 and 8.2). The dominant species, *M. coloradoensis*, accounts for 43 percent of the total fauna. Only five species comprise 75 percent of the fauna. The most abundant species in the living fauna is *N. moestus*, which occupies the fourth highest position in the death assemblage with a representation of 5 percent. Most species are rare, contributing to less than 1 percent of the total fauna. Noteworthy among the most abundant taxa are some hard-substrate species, including *Calyptrea mamillaris* (0.88 percent) and *Crepidula striolata* (0.49 percent). The presence of these species may be explained by their habit of using other shells as a substrate.

Ecological Effects of Upstream Dams and Diversion of Colorado River Water

Evidence from the shelly fauna strongly suggests dramatic changes in the benthic ecology of the Colorado River delta. Research indicates that death assemblages (thanatocoenosis) consist largely of shells that were alive before the construction of upstream dams and diversions (Kowalewski et al. 1998). Time averaging is a process by which skeletal remains are mixed at

different times and preservation is biased toward the strongest organic remains, which eliminate small and fragile forms. In this regard, the living fauna is greatly impoverished due to the sum of ecological changes and taphonomic processes in comparison with the fauna from cheniers and thanatocoenosis. There are 23 percent more species recorded on the cheniers and 36.9 percent more on the mudflat thanatocoenosis than found presently on the delta. Such large differences between the living and death assemblages can only be explained by taphonomic bias and time averaging.

An estimate of the magnitude of time averaging is indicated by the radiocarbon date of 1,050 ± 80 years BP on a modern chenier (Kowalewski et al. 1994). This means that at least 1,000 years passed during which shells accumulated and mixed together on the modern cheniers. The death assemblages are mostly composed of small shells or fragile forms with low probability to survive reworking and dissolution. They exist on the mudflats because most are buried in the mud, which increases the odds of preservation. Nevertheless, strong storm waves are capable of unearthing, transporting, and destroying the shells on the coastline where cheniers have developed.

In the delta area, only *G. palmeri* has been studied from a taphonomic perspective (Kowalewski et al. 1994). These authors found that *G. palmeri* undergoes quick mechanical disintegration after death and disappears within days or weeks after being washed out by storms and deposited on the cheniers. This condition explains why such a relatively abundant species in the living fauna is so rare (0.02 percent) in the thanatocoenosis.

From an ecological point of view, the dating of 576 ± 41 years BP of a death assemblage of the estuarine species *M. coloradoensis* (fig. 8.1) at Campo Don Abel (Téllez-Duarte et al. 2005) indicates that at that time diluted waters allowed stable populations to form 70 km away from the river mouth under conditions of strong flow. Oxygen-isotope studies also suggest that prior to the construction of Hoover Dam this species lived in brackish water (Rodríguez et al. 2001a, 2001b). In this regard, *M. coloradoensis* is a key species to understand ecological changes, because its life cycle depends on freshwater mixing with seawater. Today, this species comprises less than 1 percent of the living fauna, while it makes up an average of 43 percent of the shells in the thanatocoenosis from the same tidal mudflats (Avila-Serrano et al. 2006) and 84 percent

Table 8.2. Checklist of the shelly-fauna thanatocoenosis on the Cororado River delta.

Species	Total	%	Species	Total	%
Mulinia coloradoensis	43,996	43.22	Solenosteira capitanea	137	0.13
Tellina amianta	14,514	14.26	Epilucina californica	134	0.13
Saxicava pacifica	10,064	9.88	Anachis sanfelipensis	125	0.12
Nassarius moestus	5,105	5.01	Nassarius guymasensis	119	0.12
Chione gnidia	3,262	3.20	Trachycardium consor	110	0.11
Donax navicula	2,449	2.40	Chione pulicardia	109	0.11
Nuculana impar	2,018	1.98	Crucibulum spinosum	106	0.10
Tagelus affinis	1,525	1.50	Tellina ochraca	91	0.09
Cumingia pacifica	1,374	1.35	Crepidula uncata	80	0.08
Semele guymasensis	1,303	1.28	Cryptomya californica	77	0.08
Aligena cf. nucea	998	0.98	Tellina simulans	77	0.08
Calyptrea mamillaris	898	0.88	Cerithium stercusmuscarum	77	0.08
Nassarius gallegosi	877	0.86	Tellina ulloana	75	0.08
Anomia adamas	760	0.74	Pandora punctata	62	0.07
Macoma siliqua siliqua	742	0.73	Anomia peruviana	59	0.06
Leptopecten palmeri	726	0.71	Crucibulum pectinatum	57	0.06
Striguilla interrupta	720	0.70	Cardita laticosta	56	0.05
Mellita longifissa	682	0.67	Family Veneridae	56	0.05
Protothaca grata	665	0.65	Ostrea angelica	56	0.05
Felaniella sericata	631	0.62	Olivella zanoeta	50	0.05
Polinices recluzianus	609	0.60	Hormospira maculosa	44	0.05
Nassarius ioides	575	0.56	Cerithidea mazatlanica	41	0.04
Pitar concinnus	575	0.56	Haplocochlias sp.	41	0.04
Crepidula striolata	501	0.49	Turbonilla sp.	34	0.04
Chione sp. #1	478	0.47	Nuculana sp.	32	0.03
Ostrea palmula	448	0.44	Ensis californicus	30	0.03
Nuclulana costellata	415	0.41	Anachis nigricans	28	0.03
Chione fluctifraga	397	0.39	Trachycardium panamense	27	0.03
Cosmioconcha palmeri	390	0.38	Phos dejanira	26	0.03
Tellina meropsis	308	0.30	Terebra rufocinerea	25	0.03
Tagelus californianus	294	0.29	Encope grandis	24	0.02
Crucibulum sp.	205	0.20	Strigilla dichotoma	23	0.02
Dosinia dunkeri	198	0.19	Turritela anactor	23	0.02
Ostrea corteziensis	198	0.19	Unidentified bivalve sp.2	21	0.02
Cymatium gibbosum	193	0.19	Boreotrophon albospinosus	21	0.02
Macoma sp.	155	0.15	Chione sp. #2	20	0.02

Table 8.2. *continued*

Species	Total	%
Chione cortezi	20	0.02
Modulus disculus	20	0.02
Crepidula incurva	19	0.02
Lima orbignyi	19	0.02
Glottidia palmeri	18	0.02
Oliva spicata	17	0.02
Protothaca columbiensis	16	0.02
Crucibulum lignaium	15	0.01
Corbula nuciformis	14	0.01
Epitonium bakhanstranum	13	0.01
Unidentified bivalve sp. 3	12	0.01
Terebra armillata	12	0.01
Nassarius limacinus	11	0.01
Anatina undulata	10	0.01
Ostrea sp.	7	0.007
Conus regularis	6	0.006
Pholalidae cf *melanura*	4	0.004
Acteocina culcitella	3	0.003
Barnea subtruncata	3	0.003
Cardiomya costata	3	0.003
Crepidula sp.	3	0.003
Cyclinella singleyi	3	0.003
Unidentified gastropod sp. 2	3	0.003
Unidentified gastropod sp. 6	3	0.003
Chione mariae	3	0.003
Collisela sp.	2	0.002
Crassinella sp.	2	0.002
Unidentified gastropod sp. 5	2	0.002
Macoma undulata	2	0.002
Marinula rhoadsi	2	0.002
Modiolus rectus	2	0.002
Natica chemnitzii	2	0.002
Balcis mexicana	1	0.001
Balcis sp.	1	0.001
Unidentified gastropod sp.	1	0.001
Unidentified gastrpod sp. 3	1	0.001

Species	Total	%
Unidentified gastropod sp. 4	1	0.001
Muricanthus nigritus	1	0.001
Corbula bicarinata	1	0.001
Nucula cf. *paytensis*	1	0.001
Nucula schencki	1	0.001
Periploma planiusculum	1	0.001
Tryphomyax mexicanus	1	0.001

on the cheniers (Kowalewski et al. 1994; table 8.3). This dramatic shift in relative abundance was likely due to the environmental effects of reduced river flow to the delta and its estuary. However, the taphonomic effect of selective conservation of shells by reworking and concentration by waves on the cheniers also must be taken into account (fig. 8.5). Other abundant species, such as *T. meropsis* and *D. navicula*, are only locally present, probably resulting from successful spatfalls. On the other hand, these species could be short-lived, distributed in patches, or both. In general, the abundance and geographic distribution of the living fauna is likely the result of substrate composition, an increase in the duration of submersion, ability to withstand extremes in temperature, and capability to feed during the tide cycles (Avila-Serrano et al. 2006).

Based on the fauna collected from cheniers, Kowalewski et al. (2000) estimated a density of 25 to 50 individuals/m² prior to the construction of upstream dams and river-water diversion. Estimates of the living fauna by Avila-Serrano et al. (2006) did not include counts of small and fragile species. This gives estimates similar to those of Kowalewski (1995), but shows a density of only 3 individuals/m². Thus, since the upstream diversion of Colorado River water began in the early twentieth century, densities of comparable-sized shelly invertebrates have declined from 25–50 to 3–7 individuals/m², representing a reduction of 72 percent (from 25 to 7 per m²) to 94 percent (from 50 to 3 per m²) according to Avila-Serrano et al. (2006).

Salinity changes constitute another factor that may have caused a decline in overall density by reducing the population size of the once-dominant *M. coloradoensis*. The collapse in *M. coloradoensis* populations occurred despite the fact that productivity in

Table 8.3. Checklist of the Chenier shelly fauna (after Kowalewski, 2005).

Species	Total	%
Mulinia coloradoensis	5,933.5	91.06
Chione fluctifraga	242.5	3.72
Tagelus affinis	92	1.40
Anomia adamas	34.5	0.53
Ostrea sp.	28.5	0.43
Polinices spp.	27	0.41
Calyptrea spp.	21	0.32
Argopecten spp.	20	0.30
Chione gnidia	20	0.56
Cymantium gibbosum	20	0.31
Mytella guayanensis	19	0.31
Crucibulum spp.	16	0.29
Protothaca grata	12	0.25
Solenosteira sp.	5	0.18
Crepidula excavata	4	0.08
Unidentified bivalves	3	0.07
Trachycardium senticosum	2	0.06
Pitar concinnus	2	0.04
Cerithidea mazatlanica	2	0.04
Crepidula sp.	2	0.04
Chione cortezi	1	0.04
Cyclinella sacata	1	0.015
Lima hemphilli	1	0.015
Calliostoma palmeri	1	0.015
Fusinus depetithouars	1	0.015
Muricantus nigritus	1	0.015
Turritella sp.	1	0.015
Trachycardium sp.	0.5	0.007
Chione pulicardia	0.5	0.007
Tellina ochracea	0.5	0.007
Tellina simulans	0.5	0.007
Felaniella sericata	0.5	0.007

the northern gulf without freshwater flow remains high and appears to be limited by light rather than the availability of nutrients (Hernández-Ayón et al. 1993; Millán-Nuñez et al. 1999). Another consequence of the lack of sediment input due to upstream dams is the erosion and reworking of sediments in the area (Carriquiry and Sánchez 1999). This change in sedimentary regime (possibly related to differences in salinity) may have altered the properties of the substrate and influenced the actual separation of the shelly fauna living in the southern delta from that in the northern delta. At Campo Don Abel, substrates consist of fine sand and silt, while the more northerly localities are dominated by mud (Thompson 1968). Diversity is higher in the southern delta, perhaps as a consequence of the coarser grain size, near-normal salinity, and less extensive tidal flats, which results in less air exposure and environmental stress for the organisms compared with the more northern area.

Even during prehistoric times, diluted waters may have been an important factor in the support of local coastal mollusk fisheries. Oxygen isotopes from archeological shell middens dated to 1025 ± 90 years BP (Téllez-Duarte et al. 2001a, 2001b) suggest a freshwater influence as far as San Felipe (fig. 8.2), on the southern limits of the delta some 70 km south of the river mouth. This observation agrees with recent salinity measurements after a controlled water release in March and April 1993 (Lavín and Sanchez 1999) and the extent of the former estuary model proposed by Carbajal et al. (1997).

Figure 8.5. Modern coastal beach shell ridge (chenier) on the Colorado River delta. The dominant species is the endangered delta clam *Mulinia coloradoensis*. Photo credit: Miguel A. Téllez-Duarte

Human activities began to interfere with Colorado River flow in 1905, when a dike broke due to heavy flooding and the river was diverted into the Salton Sea (Sykes 1937). Normal flow was re-established shortly afterward, but since 1935 most of the water and sediment from the Colorado River has been trapped behind Hoover Dam and the intertidal mudflats of the Colorado delta have ceased to grow seaward. The Colorado River also underwent many natural diversions during the past 4,000 years. Ancient Lake Cahuilla (Blake 1907) was periodically formed by river diversion into the Salton Trough (Waters 1983). At least four highstands in the lake's water level occurred during the past 2,300 years. From 1250 years BP until just prior to 400 years BP, they were closely spaced in time. Whenever flow from the Colorado River was diverted, waves and currents in the Gulf of California eroded fine sediments and concentrated the coarse shelly material into cheniers along the coast. Cheniers are chronologically ordered, with older ones located inland, and they range in age from 4830 to 2190 years BP, with modern examples formed during approximately the last 70 years (Thompson 1968).

Summary

Despite likely taphonomic bias, a comparison of death assemblages on recent intertidal mudflats with the living shelly-invertebrate macrofauna of the Colorado River delta tidal flats shows the living fauna to be much less diverse and abundant. The once-dominant estuarine mollusk *M. coloradoensis* is an important key to investigating the scope of alteration to the delta ecosystem. More research is needed, however, to understand the ecological and taphonomic relationships of the diverse shelly thanatocoenosis of the intertidal mudflats. An increase in the salinity of the northern Gulf of California and changes in sedimentation regime due to the cessation of Colorado River flow are likely the main causes of ecological changes during the last century.

9 Growth of Pliocene-Pleistocene Clam Banks (Mollusca, Bivalvia) and Related Tectonic Constraints in the Gulf of California

Markes E. Johnson, Jorge Ledesma-Vázquez, and Astrid Y. Montiel-Boehringer

Based on the checklist compiled by Hendrickx et al. (2005), the most diverse marine invertebrates in the Gulf of California today are numbered among bivalve and gastropod species from the Phylum Mollusca. Gastropods (1,525 species) outnumber bivalves (565 species) by somewhat less than 3:1. Population density tells another story much in favor of the prolific bivalves that dominate intertidal and shallow subtidal clam banks. Far less is known about mollusk diversity based on Pliocene and Pleistocene fossils from the Baja California peninsula and associated gulf islands. The classic reference to the region's invertebrate paleontology remains that by Durham (1950), who provided taxonomic descriptions and illustrations for 126 bivalves and 116 gastropods. A later study by Piazza and Robba (1998) in the Loreto area revised some of the nomenclature applied to these fossils and gave more emphasis to ecological considerations.

Among the fossil bivalves described by Durham (1950), 25 percent are extinct. Pectens (or scallops) are the most common among the extinct bivalves, reflecting the group's rapid evolutionary development. More recent research by Smith (1991a) provided a wider geographic overview of Cenozoic pectens from the Caribbean to the Californias with particular reference to their biostratigraphy. Smith (1991b) also produced a companion biostratigraphic review that includes a more general account of additional bivalves and some gastropods from the gulf region. These contributions are useful for placing the various rock formations exposed throughout the Gulf of California into a temporal context that outlines the region's historical geology.

Fossil bivalves are a nearly ubiquitous component of Pliocene and Pleistocene strata in the Gulf of California. Under some scenarios, disarticulated shells were transported from the seabed where they once lived to other places where burial occurred. The Pliocene Loreto basin is an example that entailed mixed carbonate and siliciclastic sedimentation, where turbidity currents on a relatively steep slope promoted debris flows that swept away bivalves and other marine invertebrates (Dorsey and Kidwell 1999). More commonly, enormous concentrations of articulated fossil bivalves are preserved intact or only moderately disturbed from their life conditions. Clams most abundantly found in this situation belong to only a few bivalve families, such as the Ostreidae (oysters), Pinnidae (pen shells), Veneridae (Venus shells), Glycymeridae (bittersweet shells), and the Pectinidae (scallops). Members of these families typically are gregarious in habit and capable of forming massive limestone deposits.

The aims of this chapter are three-fold: (1) to describe some of the areas where extensive limestone deposits are composed of fossil bivalves primarily belonging to a few families; (2) to show how satellite imagery captures the large scale on which such deposits were formed; and (3) to demonstrate how clam-bank stability over time was affected by tectonic factors in the Gulf of California. It is worth noting that some bivalves, such as the pearl oyster among others, have disappeared from the gulf in recent times (Sáenz-Arroyo et al. 2006). The fossil record is the only lasting memorial to their former extensive distribution.

Definitions and Habitat Parameters

Stanley (1968) summarized the evolutionary history of the Bivalvia using a series of drawings that graphically capture the wide array of life habits adopted by members of this class. Bivalves are mainly suspension feeders, specialized for removing small food particles floating in the water column using sophisticated pumps and filters. Epifaunal suspension feeders include the oyster, which rests on the seafloor or commonly attaches itself to other individuals of its kind.

Infaunal nonsiphonate suspension feeders are characterized by the pen shell, which lies buried in sandy sediment with enough of its shell margin protruding above the sediment surface to permit direct access to water. This habit also is known as a semi-infaunal lifestyle. Infaunal siphon feeders include many smooth-shelled species, like the Venus clam, which live fully buried in mud or sand and maintain a connection to the surface by way of a fleshy tube (or pair of tubes) called a siphon. Most epifaunal bivalves are cemented to the seafloor or fixed in place by tough strands of organic material called a byssus. The pectens, however, are seminektonic due to their slightly asymmetrical valves, which afford hydrodynamic lift and movement through the water over short distances when the strong adductor muscle snaps the valves closed.

Based on changes in the number of genera represented by fossils from different time intervals, Stanley (1968) demonstrated that species diversity in epifaunal suspension feeders and infaunal, nonsiphonate suspension feeders expanded during the Mesozoic but suffered a decline during the ensuing Cenozoic. In contrast, infaunal siphon feeders were poorly represented during the Mesozoic, but experienced a dramatic expansion in species diversity during the Cenozoic. All bivalve lifestyles, including epifaunal, semi-infaunal, infaunal, and seminektonic are well represented in the Pliocene and Pleistocene fossil record from the Gulf of California. Although members of the oyster family are derived from much older stock than the venerids, for example, huge populations from both thrived in the gulf.

Studying Large-Scale Fossil Clam Banks

The same LANDSAT and Advanced Spaceborne Thermal Emission and Reflection Radiometer (ASTER) satellite images that proved to be useful in tracking former shorelines around the Gulf of California (chap. 2) were employed in the systematic search for shelly marine limestone. For ETM+ data, bands 7,4,1 were combined to create composite images that showcase limestone deposits. Bands 8,3,1 proved to be the best three-band combination when using ASTER data. Although it is not easy to distinguish dune rocks (chap. 10) from shelly marine limestone using these images alone, the composition of shelly marine limestone tends to have a mauve to light tan coloration that can be followed colaterally from place to place and stands out particularly well in fault blocks that are offset

from one another. Particularly in the northern state of Baja California, the geological map by Gastil et al. (1973) was a useful guide to potential spots where fossil clam banks might be explored.

Because one of the goals of this study is to appraise the population size of particular species from large-scale fossil clam banks, it was necessary to collect data on population density within a stratigraphic context tied to sites identified by satellite image. Census data were collected using a 50-×-50-cm quadrat. Identifications were keyed with reference to the taxonomic treatments by Durham (1950) and Keen (1971) using mainly field photos. This review describes the major habitat groups for bivalves, summarizing relationships found at significant localities where Pliocene-Pleistocene deposits occur.

Epifaunal Oyster Banks
San Francisquito Pliocene

The 10-km^2 San Francisquito basin (SF in fig. 9.1A; see also chap. 3) was eroded in the Middle Pliocene from granodiorite basement rocks. Irregularities in the hard rock on the floor of the basin served as appropriate places for the settlement and attachment of oysters during the earliest stages of flooding. A mounded oyster colony with a diameter of about 10 m is dissected by an arroyo in the east-central part of the basin (see basin map in fig. 3.6 for location). Maximum thickness of the mound is about 1.5 m. A single species, the coarsely plicated *Ostrea fischeri*, forms a dense aggregation of articulated individuals. About 23 whole oysters fill the vertical space (outcrop cross section) in a 50-×-50-cm quadrat. Maximum shell length was measured as 8.7 cm, and the minimum length was 3.2 cm. Maximum shell width was determined as 7.5 cm, and minimum width was 3 cm. Based on a sample of 100 articulated individuals for this species, the average shell length was found to be 5.1 cm and the average shell width was 4.9 cm. Shell length was slightly to moderately longer than shell width in 52 percent of the individuals sampled, but measurements were equal in 18 percent of the sample.

Las Barracas to El Rincón Pliocene

Mesa Las Barracas is a plateau 6 km^2 in area that overlooks Bahía Santa Inés south of the Punta Chivato promontory (fig. 9.1B). The plateau is formed by layers of Upper Pliocene limestone that dip 6° to

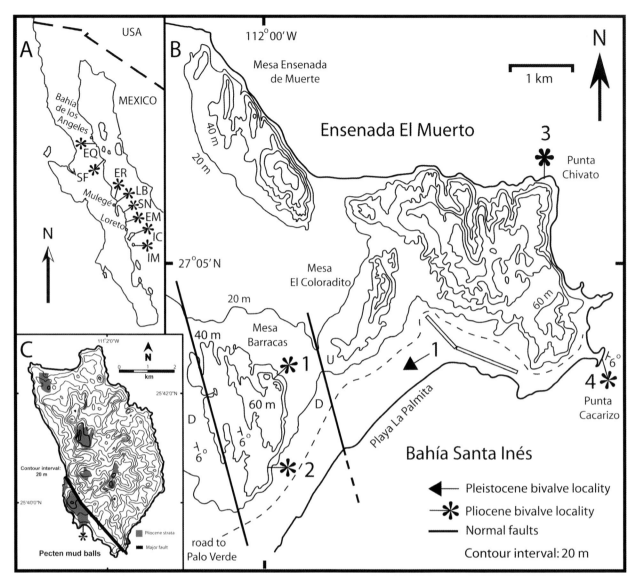

Figure 9.1. (A) The Baja California peninsula, with asterisks marking the locations of Ensenada El Quemado (EQ), San Francisquito (SF), El Rincón (ER), Mesa Las Barracas (LB), San Nicolás (SN), El Mangle (EM), Isla Carmen (IC), and Isla Monserrat (IM); (B) complete topographic coverage for the Punta Chivato region with numbered localities for Pleistocene-Pliocene bivalve localities; and (C) inset map of Isla Monserrat showing the location of strata rich in Pliocene pectens in the southwestern part of the island.

the southwest and are sliced by a series of normal *en echelon* faults that created a succession of linear escarpments oriented close to N20°W. The place name "Las Barracas" was adopted by Anderson (1950) for the repetitious slanted-roof appearance of the hills, looking like a row of barracks viewed in profile from the coast. A major fault between mesas Las Barracas and El Coloradito is traced by the impingement of a low saddle in the intervening topography that is in alignment with an offset on the coast at the southern end of Playa La Palmita (fig. 9.1B). To the east, the

up-thrown side of the fault reveals older limestone from the Lower Pliocene generally exposed in unconformable contact with Miocene basement rocks around the Punta Chivato promontory (Simian and Johnson 1997; Johnson 2002a, 2002b). It is the age difference between limestone formations juxtaposed on opposite sides of this fault that distinguishes it as a major lineament.

Exposed on the eastern rim (fig. 9.1B, Pliocene locality 1), the highest part of Las Barracas is formed by resistant cap rock constructed entirely of fossil oysters having a maximum thickness of 0.75 m. The rim is reached by a steep climb through 36 m of limey siltstone enriched in small fossil pectens (*Argopecten circularis* and *Nodipecten arthriticus*) that weather out from the poorly cemented strata immediately below the cap rock (upper part of unit 5 in the stratigraphic profile shown in fig. 9.2). Cut out from below by the erosion of the softer siltstone, large blocks of the cap rock have tumbled down the escarpment and fragmented such that great numbers of loose oysters lie scattered on the low flats across the eastern approach to Las Barracas. Individual oysters are commonly 12 cm in length, but also feature an unusual thickness (up to 10 cm) that accreted through as many as 35 to 40 distinct layers on the lower valve (Johnson 2002a, fig. 23). Because today a full-grown oyster of about the same shell length reaches maturity in only 10 years (Möbius 1883), it is unlikely that every growth line on these unusually thick fossils represents a year. The cap rock at the summit of Las Barracks is limited in extent, covering only a small part of the cuesta to the west.

The same type of cap rock thinly blankets several fault blocks situated west and northwest of Mesa Las Barracas, stretching beyond San Marcos Tierra to the Estero San Marcos on the northern coast of the Punta Chivato promontory. The surrounding region, which covers approximately 130 km², is superbly shown by a

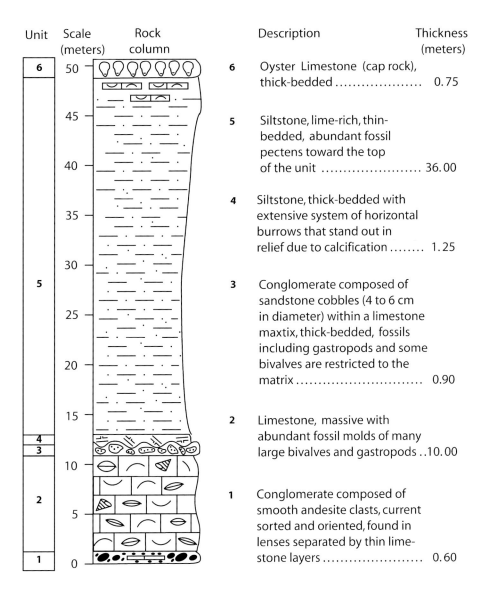

Figure 9.2. Stratigraphic column for the Upper Pliocene succession at Mesa Las Barracas, with emphasis on layers rich in fossil bivalves.

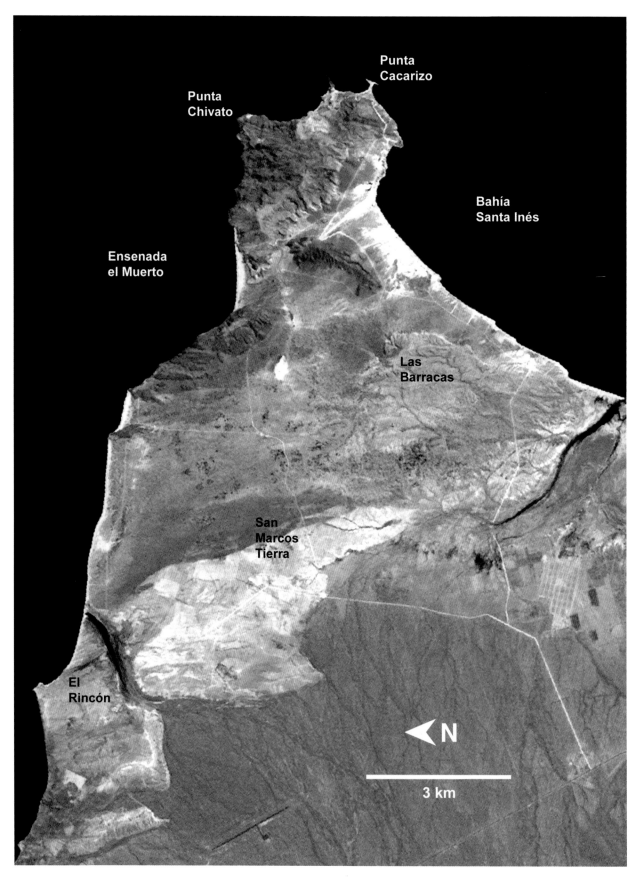

Figure 9.3. Satellite image for the region covering the Punta Chivato promontory west to Estero San Marcos and south to Mesa Las Barracas. Data set was collected by NASA LANDSAT Program, LANDSAT TM scene LT5036041009516110, Geocover, USGS, Sioux Falls, June 10, 1995.

satellite image that highlights the juxtaposition of the various fault blocks in pale mauve to ash-gray tones in contrast to the surrounding brown-gray alluvium (fig. 9.3). The village of Palo Verde at the junction of Mexican Federal Highway 1 and the Punta Chivato road is clearly visible toward the bottom of the image, as is the paved landing strip (1.7 km in length) that serves the town of Santa Rosalía (out of view to the north). With an area of about 9 km², the largest rectangular block is distinctly seen less than 2 km east of the landing strip. The terrain on these blocks is at a lower elevation and much less dissected by arroyos than Mesa Las Barracas.

Figure 9.4A shows strata from a small fault block near El Rincón cut by a shallow arroyo that runs a short distance to the adjoining alluvial plain with its extensive *cardonal* (cactus forest). Seen before the uneven skyline marked by the Sierra de Guadalupe in the far distance is the level fault scarp of another limestone block. As at Las Barracas (fig. 9.2, unit 5), the most characteristic strata within these blocks consist of light tan, limey siltstone (fig. 9.4A). In this case,

the cap rock on the El Rincón block is less than 0.5 m in thickness, but is crowded with large, saucer-shaped oysters having a maximum shell length of 15 cm (fig. 9.4B). For the most part, the oysters are articulated and preserved in their original growth position. The monospecific population is attributed to *Ostrea cumingiana*. On average, 12 recumbent oysters fill a 50-×-50-cm quadrat, which works out to 48 oysters/m² of substrate. Thus, 1 km² of oyster cap rock represents a piece of former seabed with about 48 million oysters. Much of the cap rock has been eroded away from the areas around Las Barracas, San Marcos Tierra, and El Rincón to expose the underlying siltstone, but no more than a single marker bed with abundant oysters has been found in strata from any of the Upper Pliocene fault blocks in the region. Assuming that the cap rock from different blocks was deposited as a coeval entity, the total area could be 20 km². Not taking into account the actual volume of the cap rock, the surface area represented by a 20-km² seabed could potentially entail a population approaching a billion oysters.

Figure 9.4. (A) Near El Rincón looking northwest across fault blocks of Upper Pliocene strata capped by an oyster limestone (40 cm in thickness) with the Sierra de Guadalupe on the skyline; and (B) a large fossil (*Ostrea cumingiana*) from the cap rock (9-cm pocket knife for scale). Photo credits: Markes E. Johnson

Playa La Palmita Pleistocene

Incipient oyster mounds developed near the center of a small basin (1.5 km²) behind Playa La Palmita on Bahía Santa Inés south of the Punta Chivato promontory (fig. 9.1B, Pleistocene locality 1). Much of the carbonate material accumulated there is a loosely consolidated mass of shells, sand dollars, and extensive rhodolith debris, but the oyster mounds are better cemented as distinct entities. Virtually all oysters within the mounds are articulated and preserved in growth position. The largest mound, 30 m in length and 40 cm in thickness, is dissected by one of the arroyos traversing the basin. Low, ridge-like irregularities in Pliocene sandstone on the floor of the Pleistocene basin became the locus of oyster colonization during a rise in sea level. The ridges are oriented perpendicular to the present shores of Bahía Santa Inés, possibly as a result of tidal scour. Viewed in cross section on the sidewall of the arroyo, successive layers of oysters are stacked on top of the pioneering population that first colonized the sandstone ridge. Because the mound does not appear in the opposite wall of the arroyo, it is impossible to estimate the width of the structure, but thousands of individual oysters are preserved in the lone embankment. Allowing for no more than a dozen years for each generation in the structure, it can be estimated that the mound grew upward over a 50-year interval (Johnson 2002a, p. 24). The populations that constructed the mound are monospecific, all probably belonging to *O. cumingiana*.

Epifaunal Bittersweet-Shell Beds
Punta Chivato Pliocene

Sloping 8° north off underlying andesite basement rocks into the Gulf of California, ramp deposits composed of basal conglomerate and limestone beds are well developed near Punta Chivato (fig. 9.1B, Pliocene locality 3). Anderson (1950) and Durham (1950) attributed these strata to the San Marcos Formation in the Lower Pliocene, based partly on occurrence of the heavy echinoid *Clypeaster bowersi*. Simian and Johnson (1997, fig. 4) illustrated the fauna from this locality, which except for the echinoids is represented exclusively by fossil molds. Limestone layers immediately above the basal conglomerate are particularly rich in the small bivalve *Glycymeris maculata*, which attains a shell size typically 5 cm in length. Unlike its larger, deeper-water cousin, *Glycymeris gigantea*, this

species is mainly intertidal in habit. It thrives on sandy surfaces, often nestled in pockets among coastal boulders. All fossil bivalves from this locality are articulated, which implies rapid burial.

Ensenada El Quemado Pliocene

The northernmost locality in the gulf with a significant limestone deposit that preserves a Pliocene setting with abundant fossil bivalves occurs at Ensenada El Quemado, 12 km east of Bahía de Los Angeles (fig. 9.1A). Mapping by Gastil et al. (1973) delineated geological relationships around the ominously named Punta Que Malo that show the 4-km² promontory was cut off from the peninsular mainland as a paleoisland during the Pliocene. The promontory is formed by quartz dacite of Cretaceous age that was later intruded by basalt. A 250-m-wide channel filled with Upper Pliocene limestone runs for approximately 2 km from Bahía San Juan on the eastern side of the promontory to the opposite side, opening onto Ensenada El Quemado to the east with a broader expanse of limestone. Contact is exposed with basement rocks that formed the bottom of the channel, and some arroyo sections show the accumulation of a basal conglomerate with cobbles of quartz dacite mixed with minor amounts of metamorphic rocks introduced from a neighboring fault boundary to the south. The maximum thickness of the conglomerate is 1.25 m and, although the cobbles are suspended in coarse sediment, fossils are not plentiful therein.

Above the basal conglomerate, the limestone remains heavily impregnated with large granules of quartz (1.5 to 2 mm in diameter) eroded from parent diorite rocks. Maximum thickness of the channel deposit above the basal conglomerate is 5.5 m and a Upper Pliocene position is confirmed by the presence of the large echinoid *Clypeaster marquerensis*. Various bivalve zones of domination are evident within the channel. Disarticulated deposits of *Ostrea fisheri* and *O. megadon* are restricted more toward the opening of the channel at El Quemado. Farther into the channel, however, the epifaunal *G. gigantea* becomes the primary faunal element. These bivalves, which reach a maximum shell length of 10 cm, are usually found articulated and dispersed evenly throughout the channel beds with a density of about 3 individuals/m². According to Keen (1971, p. 55), dredge hauls in water depths between 7 and 13 m commonly collect this

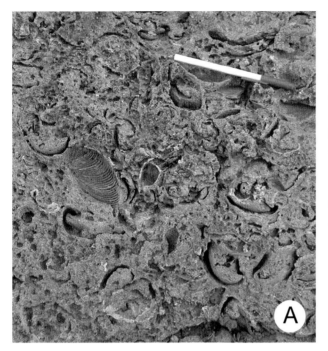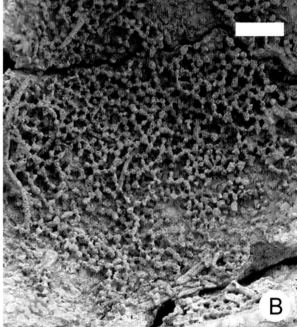

Figure 9.5. (A) Mesa Las Barracas (see fig. 9.1B for location), showing outcrop of massive limestone (15-cm pen for scale) with abundant infaunal bivalves preserved as fossil molds; and (B) detail from the internal mold of a bivalve showing extensive shell borings (scale = 7.5 mm) belonging to the trace fossil *Entobia lacquea*. Photo credits: Markes E. Johnson

species. The paleoecological implications for the Pliocene deposit at El Quemado are that the channel was not very shallow and that the large bittersweet shells were buried below the surface fairly rapidly before disarticulation occurred.

Infaunal Clam Banks
Las Barracas Pliocene

A massive, 10-m thick embankment of limestone with a high concentration of fossil bivalves dominated by infaunal species is exposed on the southeastern side of Mesa Las Barracas (fig. 9.1, Pliocene locality 2), presumably from a part low in the stratigraphic column (fig. 9.2, unit 2). Much in contrast to the Barracas oyster bank described previously, only a small percentage of the bivalves from this interval are intact as articulated shells. Moreover, the original shells were removed by dissolution so that the fossils occur as molds. External molds show growth lines etched on the outside surface of shells no longer present. Fine details of shell growth are preserved by sediment that pressed against the shell exterior prior to its dissolution (fig. 9.5A, left-center). Internal molds result from packing of sediment within the shells to capture features such as muscle scars embedded on the

inner surface. In cross section, arc-shaped cracks in the outcrop where shells once existed are up to 15 cm in length. This indicates that the shells were large and that minimum breakage occurred. Based on recovery of internal molds from articulated shells, the most common species are *Megapitaria squalida* (chocolate clam) and *Laevicardium elatum* (heart cockle). Both are strong burrowers found living today buried in intertidal to shallow-water sand flats. Other less common molds from this locality are identified as *Arca pacifica* and *Anadara grandis* (small and large arc shells) or *G. gigantea* (bittersweet shell), all epifaunal elements. The mixture of species, some surface dwellers and others subsurface dwellers, combined with the absence of distinct layering in the limestone shows that the entombing sediments on the sea bottom were thoroughly mixed by burrowing organisms (or bioturbated) prior to lithification.

Additional evidence that shells on the seafloor were plowed under for burial with the rest is demonstrated by the delicate pattern preserved in bas-relief from the inner, curved surface of a large valve (fig. 9.5B). The lacey, web-like pattern is the complex result of a process that began with a clionid sponge colonizing a clamshell that lay exposed on the seabed. These endolithic sponges have the habit of boring into shells, corals, or limestone to create chambers where they can live safely anchored. Connected networks of chambers are etched into calcium carbonate using mild hydrochloric acid secreted by the sponges. It is all the same to the sponge if the host coral or clam is alive or dead. Although firmly embedded, the sponge must have access to the surface to filter water for food. In-filling of the small chambers by fine sediment after the sponge colony died led to preservation of a trace fossil, the likeness of which we attribute to *Entobia lacquea* (see Bromley and D'Alessandro 1984, pp. 244–246). The network of former cavities was exposed in bas-relief when the shell was dissolved, leaving the fillings intact.

Above the massive shell embankment at the same locality on Las Barracas is one of the most unusual rock formations we have encountered anywhere in the Gulf of California (fig. 9.2, unit 3). It consists of conglomerate formed by cobbles that typically measure 4 × 6 × 10 cm and are very lumpy in shape. These clasts are densely crowded together, weather orange in color, and on first impression appear to be fine-grained sandstone (fig. 9.6A). Thickness is irregular, but the stratum is always less than 1 m thick. A count of 400 grains from each of two thin-section trials revealed the clasts to be dominated by carbonate microspar

(58 ± 5.0 percent by grain count) and volcanic grains consisting of small, nested plagioclase twins (24 ± 4.5 percent). Larger plagioclase laths (fig. 9.6B, up to 0.65 mm in length) account for 13 ± 3.5 percent of the total count. Trace amounts of clinopyroxene, biotite, and hornblend make up the balance. No macrofossils occur within the cobbles, but rare relict grains of carbonate appear to be derived from bivalve shells and echinoid platelets.

The conglomerate is well cemented by a limestone matrix with abundant molds of macrofossils preserved in the interstices of the cobbles. Gastropods such as *Strombus granulatus*, *Conus arcuatus*, and *Oliva* cf. *kaleontina* are represented. Pockets of limestone at the top of the bed are bored by *Lithophaga plumula* (fig. 9.6C), a short pencil-shaped bivalve that, like clionid sponges, uses dilute acid to etch its way into any hard substrate composed principally of calcium carbonate. In some cases, lithophagid bivalves also penetrated the "sandstone" cobbles.

No parent strata survive anywhere in the surrounding region to which the peculiar clasts that form the conglomerate can be traced. Emplacement by a mass gravity slump is a logical scenario that obviates the source, takes into account the atypical "lumpy" nature of the cobbles, and explains the deposit's irregular

Figure 9.6. (A) Mesa Las Barracas (see fig. 9.1B, Pliocene locality 2), showing sandstone-eroded conglomerate (orange) set in carbonate matrix (gray), with 9-cm pocket knife for scale; (B) a petrographic thin section from one of the large clasts, showing abundant grains of the volcanically derived minerals plagioclase (striped) and hornblend (dark); largest bits up to 0.65 mm in length; and (C) a boring bivalve (*Lithophaga plumula*) from top part of the outcrop. Photo credits: Markes E. Johnson

thickness (Johnson et al. 2002). Based on variations in plagioclase grains found within the clasts, at least two separate volcanic events were involved. Delivery of the plagioclase and other volcanic grains was by way of wind-blown silt eroded from different andesite flows. The silt may have been dropped in the shore face on the sheltered side of nearby islands where fine carbonate sand and mud constituted the primary coastal facies. As lithification proceeded, most of the original biogenic grains and micritic mud was replaced by sparry calcite, leaving volcanically derived minerals and lithic grains unaltered. Prior to complete lithification, a minor earthquake struck the shallows and sent semilithified clots of sediment into deeper, adjacent waters as a kind of submarine avalanche. Based on the depth distributions of extant species matched to the fossils found in the conglomerate matrix (Keen 1971), the slide came to rest in less than 50 m of water. Lithophagid bivalves found a suitable environment on which to settle into the consolidated rubble at the top of the deposit.

El Rincón Pleistocene

An unconsolidated mass of Pleistocene shells 0.5 m in thickness and covering an oval-shaped patch of ground approximately 250 m² in area is cut by the dirt track that runs from Mexican Federal Highway 1 to El Rincón (fig. 9.3). The intertidal to shallow-water origin of the deposit is hinted at by the presence of the gastropod *Turbo fluctuosus*, but otherwise bivalves heavily dominate the assemblage. Some bivalves remain articulated, but due to the loose sand in which they occur, the edge of the graded road is littered with disarticulated shells. The infaunal bivalve *M. squalida* is the most abundant species present, with other components supplied by *Chione californiensis*, *Chione tumens*, and *Tagelus affinis*.

Semi-Infaunal Pen-Shell Beds
San Nicolás Pliocene

Basal limestone from a ramp unconformity in the San Nicolás basin (fig. 9.1A) is crowded with the fossil molds of whole pen shells (Johnson and Ledesma-Vázquez 2001, fig. 6B). Individuals are articulated but stacked horizontally near the unconformity with underlying Miocene andesite. Unlike most other Miocene-Pliocene boundaries in the gulf region, the contact is oddly devoid of clasts eroded from basement rocks.

This suggests that the unconformity surface, which follows an uneven 6° slope, was swept clean. The stratigraphic context is inconsistent with the habits of pen shells, known to live partially buried vertically in mud or sand in protected bays (Keen 1971, p. 74). Furthermore, pen shells possess relatively thin and brittle shells. We conclude that the Pliocene shells were scoured from a nearby locality and rapidly deposited as a large mass of material subsequently cosseted from any further disturbances.

Individual molds from the San Nicolás locality are large and with a maximum body length of 19 cm and width of 11 cm. These dimensions compare well with mature individuals of *Pinna rugosa* observed by the authors living in intertidal sands on the sheltered side of Isla El Requesón in Bahía Concepción. The fossils have not been determined as to species, but they share affinities with *P. rugosa*.

Isla Coronados Pleistocene

The southern side of the small volcano on Isla Coronados (in view of Loreto, fig. 9.1A) shelters Pleistocene lagoon deposits composed of extensive rhodolith debris that accumulated to a thickness of 6 m (see chap. 7). A fixed barrier of tiny andesite knolls also confines the paleolagoon on its outer margin, and the uniformly north-dipping rhodolith limestone within the lagoon is interpreted as an over-wash deposit (Ledesma-Vázquez et al. 2007a). Prior to a rise in sea level and burial by rhodolith debris, a cluster of *Pinna corteziana* colonized the area behind the opening between two of the andesite knolls. The occurrence was registered by Durham (1950) and confirmed by Ledesma-Vázquez et al. (2007a). Larger populations may have occupied the site, but only a small part of the lagoon bottom at the appropriate stratigraphic horizon has been excavated by natural erosion. We observed individual shells at this locality with a maximum length of 25 cm and width of 8.5 cm. The paleoenvironmental setting is fully consistent with that of living pen shells, as for example in Bahía Concepción.

Scallop Banks
Punta Cacarizo Pliocene

At the end of a tombolo southeast of the Chivato Promontory (fig. 9.3) is the strangely shaped Punta Cacarizo (also locally known by American residents as Hammerhead Point). The blunt, 320-m wide snout of the point is a geological outlier (fig. 9.1B, Pliocene locality 4) related to other outcrops on the promontory. Nearly 5 m of basal conglomerate and overlying limestone mark the position of a former Pliocene shoreline (see chap. 3) and dip 6° to the southeast into the open gulf. The upper layers of thick-bedded limestone on Punta Cacarizo are crowded with external molds of abundant fossil pectens. Unfortunately, karst erosion on the limestone and general incompleteness of the molds make species identification difficult. The site does, however, represent a good example of a scallop bank that developed in shallow coastal waters off east-facing shores during the Early Pliocene (San Marcos Formation). Layering of pecten beds across the point runs parallel to the former shoreline.

El Mangle Pliocene

Another example of Pliocene pecten beds that accumulated along an east-facing paleoshore occurs at El Mangle, 30 km north of Loreto (fig. 9.1A). Described by Johnson et al. (2003), the site is unusual for its preservation of a wedge-shaped package of transgressive-regressive strata that terminate against a rocky shoreline formed by Miocene andesite. The scallop beds, which have a maximum thickness of about 12 m, are dominated by *Argopecten abietis*. Shell material is preserved and includes encrustation by barnacles. A Middle Pliocene position for this deposit is based on the dating of an associated tuff unit that yields an age of 3.3 ± 0.5 Ma. Regarding spatial relationships, the most compelling aspect of these pecten beds is their location in the thickest part of the limestone wedge at a distance that can be accurately measured as more than 300 m distal from the paleoshore. An entirely different fauna is preserved in the laterally thinning limestone adjacent to the paleoshore.

Arroyo Blanco Pliocene

Yet a third example of Pliocene scallop beds that accrued along a former land area is exposed in the canyons of Arroyo Blanco on the eastern side of Isla Carmen (fig. 9.1A). Eros et al. (2006) determined that limestone layers rich in fossil pectens account for 20 percent by volume of the Arroyo Blanco basin, which covers an area of 3.87 km². A significant aspect of this basin is the unusual span of time over which rhodolith beds and pecten beds were deposited during most of the Pliocene and part of the Pleistocene. This

relationship is underscored by the large number of pecten species recorded from different stratigraphic intervals within a 60-m-thick succession. The various species include *Argopecten abietis, Argopecten antonitaensis, A. circularis, Argopecten sverdrupi, Lyropecten modulatus, Patinopecten bakeri* subsp. *diazi,* and *Pecten vogdesi*. Details regarding basin tectonics for this locality are given in chapter 3.

Isla Monserrat Pliocene

Limestone rich in fossil pectens dominated by *A. abietis,* but also including *Lyropecten subnodosus* and *P. bakeri* subsp. *diazi,* covers about 60 ha in the southwestern part of Isla Monserrat (figs. 9.1A, C). According to Durham (1950), *P. bakeri* subsp. *diazi* (fig. 9.7A) is diagnostic for the Middle Pliocene in the Gulf of California. Good shell material is preserved, but almost all pectens are disarticulated. The deposit, which extends laterally over an 84-m change in elevation, follows a dip angle of 6° from north to south but the incline shallows to 3° at the outcrop's southern terminus near the island's present shoreline. A normal fault is oriented N45°W adjacent to the rectangular outcrop, separating it on the down-thrown side of a fault escarpment with the footwall formed by Miocene andesite (fig. 9.1C). Apparent weathering of andesite contemporary with growth of the pecten bank resulted in the addition of a minor amount of clay minerals to the carbonate rocks.

Features as yet unique to the gulf Pliocene that can be compared to armored mud balls (Ledesma-Vázquez et al. 2007b) are exposed in cross section in the walls and floors of arroyos crossing the lower, southern part of the outcrop area. Figure 9.7B illustrates an example eroded in bas-relief on the floor of an arroyo. The structure, which is 60 cm in diameter, is formed by a multitude of single pecten shells packed together with valves nestled inside valves in a partly radial, partly tangential orientation to the perimeter of the eroded circle. Other examples, also circular in cross section but as small as 30 cm in diameter, are exposed in arroyo walls. Circular cross sections available in two planes mean that the structures are actually spherical in shape.

Strictly defined, armored mud balls are firmly adhesive constructions of clay and sand-size particles that grow by physical accretion much like a snowball rolling down slope. Generally, they are associated with a terrestrial environment in ephemeral stream

Figure 9.7. The Middle Pliocene pecten beds on Isla Monserrat showing (A) *Patinopecten bakeri* subsp. *diazi* and (B) an armored mud ball composed of shell debris exposed in bas-relief (pen for scale = 13.5 cm). See figures 9.1A, C for location. Photo credits: (A) David H. Backus, (B) Jorge Ledesma-Vázquez

channels where the gradient is low and the climate typically arid. Clay wetted by rainfall provides a source of adhesion for the larger grains. The unusual twist to our interpretation is that a similar process operated in a marine setting and entailed small, empty pecten shells usually no more than 6 cm in diameter agglomerated by sticky clay together with other coarse debris. A model for the creation of the pecten mud balls is that they rolled down channels on the margin of the carbonate bank as a result of gravity slumps. The physical impetus for starting the process under intertidal to shallow-water conditions may have been strong tidal movement in channels on the Monserrat bank, which helped to sort and concentrate shell materials.

Tectonic Instability and Debris Slumps

Las Barracas and Isla Monserrat are two areas within the Gulf of California that were strongly affected by tectonic instabilities during the Middle Pliocene. The region surrounding Mesa Las Barracas is characterized by blocks of Miocene Comondú andesite faulted on a north–south axis and tilted to the west as a result of extensional tectonics during the Miocene. En echelon faults on Mesa Las Barracas have a more diagonal orientation and clearly postdate the area's Upper Pliocene strata, including the distinctive conglomerate

bed (fig. 9.2, unit 3) described as hosting lithophagid bivalves. En echelon faults and the earlier earthquake that precipitated the mass gravity slump at Mesa Las Barracas are constrained sufficiently in time to suggest linkage with the changeover to transtensional tectonics in the Gulf of California after about 3.5 Ma. The satellite image for the Punta Chivato region (fig. 9.3) also demonstrates the after-effect of tectonic activity that fractured and moved large rectangular blocks of limestone west and northwest of Las Barracas.

Likewise, Isla Monserrat was initially affected by an episode of Miocene faulting that fractured and tilted blocks of Miocene Comondú andesite as a result of extensional tectonics (in this case, tilted eastward). Diagonal faults on a northwest–southeast trend also are common on Isla Monserrat (see fig. 7.6 for an enlarged geological map of the island). These oblique faults postdate deposition of Middle Pliocene limestone in relatively shallow water and clearly involved tectonic forces instrumental in uplifting the entire island. Such tectonic activity is widely associated with the changeover to transtensional tectonics in the gulf, as also demonstrated at El Mangle in relation to the oblique El Coloradito Fault that offsets a Middle Pliocene scallop bank. More like Las Barracas to the north, however, rumblings of less violent earthquakes are echoed by the occurrence of pecten mud balls in Middle Pliocene strata on Isla Monserrat.

To a lesser degree, the unusual accumulation of transported (but undamaged) pen shells on the Pliocene-Miocene unconformity in the San Nicolás basin also bears some earmarks of a debris slump in shallow water. Collectively, internal stratigraphic features from Middle Pliocene limestone at Las Barracas, San Nicolás, and Isla Monserrat all hint at mass gravity flows related to earthquakes.

Summary of Major Relationships

Limestone strata of Pliocene and Pleistocene age with a prodigious component supplied by mollusk shells, especially bivalves, are widespread along the western shores of the Gulf of California. Although infaunal bivalves are a late evolutionary innovation that achieved great success during the Cenozoic (Stanley 1968), the gulf supported a full array of bivalves with different habitat preferences. Infaunal species such as *M. squalida* and *L. elatum* are well represented in Pliocene and Pleistocene deposits, but the pectens, which include several extinct species in the genus *Argopecten*, also attained tremendous success.

As survivors from an older Mesozoic stock of bivalves, oysters also clearly found ecological space for proliferation in the Gulf of California. Fossil oyster beds from the San Francisquito, El Rincón, and Las Barracas areas are extraordinary for the immense populations of individuals preserved in crowded growth position. By comparison, it is interesting to note the historical description by Möbius (1883, p. 690), regarding the North Sea *Wattenmeer* (sea-flats) off the coast of northern Germany that provided habitat for the Holstein oyster in a region no more than 100 km^2 under a water depth of 2 m during ebb tide. It is startling for the paleontologist to learn that the Holstein oyster never thrived under conditions that fostered gregarious growth. According to Möbius (1883), they lay singly on a substrate of old shells, and it was necessary for the oystermen to dredge an area of from 1 to 3 m^2 to retrieve one individual. Based on eyewitness accounts from the seventeenth century regarding pearl oysters in the Gulf of California (Sáenz-Arroyo et al. 2006, p. 139), the local species were more gregarious and lived in "bunches of twenty more or less" but spread across banks up to 6 nautical miles in diameter in about 15 m of water. Times have changed, but the great volume of Pliocene and Pleistocene limestone formed by mollusk shells in the Gulf of California gives us a window on the past. That window is even more fascinating to look through, on account of the insights provided by mapping ancient clam banks affected by tectonic events in the geological history of the Gulf of California.

10 Sand Dunes on Peninsular and Island Shores in the Gulf of California

David H. Backus and Markes E. Johnson

Mollusks, corals, and coralline algae contribute to the overall marine biodiversity of the Gulf of California but they also are significant as mass producers of skeletal debris left behind in bays, washed onto beaches, and blown inland to enrich coastal dunes along the eastern shores of Baja California. Carranza-Edwards et al. (1998) found distinct differences in the composition and provenance of beach sands in a survey of 50 locations equally divided between the Pacific and gulf coasts of the Baja California peninsula. Sand from barrier islands on the Pacific Ocean is mature, having been delivered to the coast by streams crossing wide coastal plains. Individual sand grains are well rounded. Sands along the gulf coast are considered immature, because they have not been transported very far. They tend to be trapped in pockets along a coastline with high topographic relief. Quartz and feldspar are the dominant components of Pacific beach sands, whereas gulf sands are highly enriched in calcium carbonate. Coastal waters in the Gulf of California are warmer, clearer, and more sheltered than the Pacific waters on the western side of the peninsula. These are the principal factors most likely to promote net accumulation of calcium carbonate sediments on gulf beaches (Carranza-Edwards et al. 1998, p. 271).

Research on the region's calcium carbonate dunes was introduced in a study by Ives (1959), who compared sand bodies in the Yuma and Sonoran deserts at the top of the gulf. In the Yuma area on the U.S. side of the international boundary, quartz sand is derived from sediments carried downstream by the Colorado River. The Yuma dunes are rich in quartz, but register only 10 percent calcium carbonate by volume. Dunes on the Sonoran side of the border are dominated more than 70 percent in volume by shell fragments between 0.01 and 2 mm in diameter. Formation of both dune fields is due to the same dynamic processes, namely westerly summer winds and strong northwesterly winds called the winter *norte* winds by Ives

(1959). Thus, sediment source plays a key role in the compositional difference between these dune fields. Beach deflation, which entails transfer of beach sediments to coastal dunes, is widely perpetuated by the norte winds that funnel southward down the axis of the Gulf of California most strongly during the winter months but also during much of the rest of the year (Johnson 2002a; Parés-Sierra et al. 2003). In particular, north-facing beaches located anywhere in the gulf are most prone to develop associated coastal dunes.

Worldwide, dune fields are calculated to cover about 5 percent of the Earth's surface (Thomas 1997). The technical term eolianite applies to a subset of dunes composed of carbonates found in coastal zones or on islands between 20° and 40° latitude in both hemispheres. According to the survey by Brooke (2001), the greater part of the world's eolianite deposits are in the Southern Hemisphere, mainly southern Australia and southern Africa. A warm climate conducive to marine production of skeletal calcium carbonate and the onshore winds necessary to transfer those materials inland are the essential prerequisites for the development of substantial eolianite deposits. Brooke (2001) argued that the tectonic setting of shorelines also predisposes a region to the attraction or prevention of coastal carbonate dunes. Tectonically active coasts typically exhibit a higher concentration of rocky shores with steep offshore profiles that allow little space for shallow shelves and related beaches. The prospects for occurrence of eolianites along the gulf coast of the Baja California peninsula and its related islands would seem to be unfavorable by this criterion.

Heretofore, no comprehensive survey of the gulf shores of Baja California has been attempted to access the frequency and distribution of coastal sand dunes of any kind. In this chapter we describe such a survey through the analysis of satellite imagery. Enrichment of dune sand by calcium carbonate detritus derived from marine shells and coralline algae is of particular interest in this study. The limited information on this topic for the Gulf of California is summarized and additional data are derived from sand samples collected by the authors at selected localities. Purity of dune sand with respect to calcium carbonate is predicted on a case-by-case basis with reference to geographic orientation and the composition of surrounding geological exposures subject to physical weathering. Conditions pertaining to coastal oceanography also are taken into account regarding the development of beaches contiguous to coastal dunes. We attempt to provide some measure of the potential for the sequestration of carbon bound in the detritus of organic skeletal materials and hence the carbonate component of these dune sands.

Assessing Coastal Dunes

Images from the LANDSAT and Advanced Spaceborne Thermal Emission and Reflection Radiometer (ASTER) satellite systems were the primary resources consulted for this study of coastal dunes in the Gulf of California. The coastline of the Baja Peninsula from Cabo San Lucas to the Colorado River delta, as well as most of the major islands in the Gulf of California, can be covered with only minor gaps by a series of 26 ASTER images. Each ASTER image processed and used for this project covers about 60 × 60 km of area and has a resolution of 15 m/pixel. Images created using Enhanced Thematic Mapper (ETM+) data from LANDSAT and downloaded from the Global Land Cover Facility at the University of Maryland provided additional coverage. The ETM+ image area is 180 × 185 km with a nominal resolution of 30 m/pixel. We further processed our ETM+ scenes to a resolution of 14.25 m/pixel using Gram-Schmidt image sharpening, a technique that utilizes the higher resolution panchromatic band 8 to improve the clarity of the original image. The remote sensing software ENVI ver. 4.2 was used to process all images used in this study.

Band combinations that proved best for the overall delineation of various rock types and geomorphic features along peninsular coastline were from ASTER images in bands 8,3,1 (repository on disk, this volume) and for ETM+ scenes either bands 7,4,1 or 5,3,1. The Google Earth internet-based map service provided higher resolution data for many regions within our study area, which allowed us to estimate the landward margins of sandy beaches and the amount of topographic expression on associated dunes. Coastal

dunes were identified and assigned reference numbers from south to north along the peninsula and from island to island.

Extent of ground coverage occupied by dunes, including relatively open and vegetated parts, was outlined using the measurement tool in ENVI ver. 4.2 as a preliminary step to the measurement of surface area in hectares (1 ha represents 10,000 m²). The degree of dune stabilization was expressed as a simple ratio between areas occupied by relatively open sand and heavily vegetated sand. Where possible, maximum dune thickness was recorded based on published data or available field notes. It is likely that ecologists have visited few of the remote dunes described here. Surface features from previously studied dunes were compared with all others to arrive at an estimation of maximum thickness for undocumented dunes. In general, any given dune can be boxed within a three-dimensional space partitioned by a diagonal plane extending from a zero point on the stoss side to the maximum height on the steep leeward side. Hence, the average thickness of an individual dune is roughly half the maximum height of the dune. Clearly, this process of geometric containment must be repeated for each part to arrive at a summary estimate of volume for a complex dune field. Satellite images also were searched systematically for evidence of dune migration based on the orientation of beaches, shape of dune crests, and windrow patterns in dune vegetation. In some examples, these data are corroborated by ground observations.

Coastal Dune and Dune-Field Formation

As described above, dunes accumulate where there is an available sediment source, sufficient wind to transport sand, and a place for windblown sand to accumulate. Sufficient wind in the Gulf of California is not a problem. The dominant winds for the region are the northwesterlies that blow directly down the axis of the gulf with an average azimuth of S24.5°E (Parés-Sierra et al. 2003). Most common during the winter months, these winds can blow persistently for days at a time with sustained velocities commonly in the 5–7 m/sec range, with gusts of up to 8 m/sec and higher not unusual (Bray and Robles 1991; Russell and Johnson 2000). Given the length of the Gulf of California, the potential for high wind velocities due to fetch alone are substantial for those parts of the coastline that project into the gulf (Bauer and Davidson-Arnott 2002).

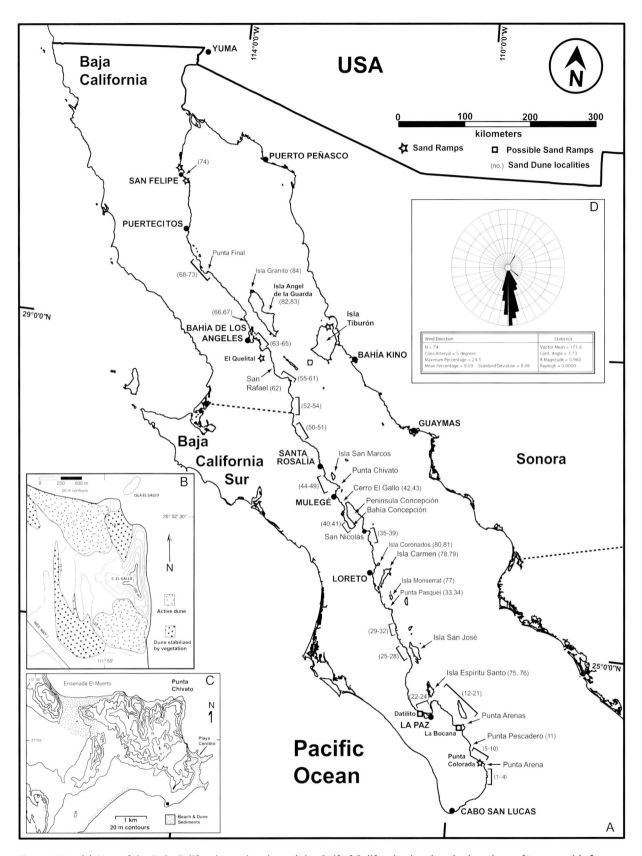

Figure 10.1. (A) Map of the Baja California peninsula and the Gulf of California showing the locations of topographic features, dunes, dune fields, and sand ramps described in the text and listed in tables 10.1–10.3; (B) map of the El Gallo dune field; (C) map of the Punta Chivato area; and (D) summary diagram for wind direction data from the gulf coast dune fields of Baja California.

Figure 10.2. View of the El Gallo dune field from Cerro El Gallo looking northwest toward the Río Mulegé estuary. Note the foredune area that sits immediately behind the beach, the lone transverse dune in the backdune area, and the older, heavily vegetated dune field in the background. Photo credit: David H. Backus

Only two rivers run persistently throughout the year from the eastern side of Baja California into the Gulf of California, the Río Mulegé and Río Colorado. Both are unusual. The Río Mulegé is a brackish-water lagoon that is only 3 km long (Roberts 1989). The discharge of the Colorado River has been severely reduced by dams and irrigation upstream so that the amount of sediment discharged currently into the gulf at the delta is marginal compared to rates in the historic and geologic past. South of San Felipe, most sediments delivered to the coast come through arroyos that are dry for months to years at a time and activated during seasonal heavy storms and *chubascos*. In the northern gulf, storms usually occur during the late autumn and winter months, while in the southern gulf, heavy rain occurs during the late summer and early autumn (Bray and Robles 1984; Douglas et al. 1993). Wind speeds during these storms are on the order of 12–20 m/sec (Bray and Robles 1984). Chubascos are tropical hurricanes that also occur during the summer and early autumn storm season. Although episodic in nature, the wind and rain associated with these storms have a cataclysmic effect on the landscape due to the sparseness of vegetation on the Baja California peninsula (Roberts 1989). Most of the peninsula is considered arid. Only the high mountain regions in the northwestern and southern

end of the peninsula that receive more than 25 cm of rain in a year are considered semi-arid. The gulf coast of the Baja California peninsula is a desert, but it receives huge amounts of water and sediment from intermountain basins whenever a storm or chubasco hits mountains upstream.

Once sediment is delivered to the coast, several possible things can happen depending on the depth and width of the local marine shelf. In areas with a low-relief coast and shallow-marine shelf, sediments brought to the coast may be stored in estuaries or passed onto the shelf, where they are reworked by winds, currents, and tide. At the other extreme, if the coast is high relief and the shelf is narrow, the sediment will bypass the shelf and be delivered into deeper water and away from the influence of coastal wind and waves. Sediments held near the coast are potential source materials for dunes. Dunes are often found associated with tidal estuaries, developing on barrier beaches and barrier islands that may with time grow large enough to cut off the estuary from the ocean (Woodroffe 2002). Many low dune areas on the peninsular coast of the Gulf of California that have flat playas behind them represent former tidal estuaries that were cut off as the local coastal systems evolved (Woodroffe 2002).

Coastal topography is quite variable over the

Figure 10.3. View from the northwest of a dune perched on the top of a north-facing cliff on the coast, south of Punta San Rafael in the state of Baja California. Isla San Lorenzo is in the background. Photo credit: David H. Backus

length of the Baja California peninsula, for example, ranging from the broad, flat coastal plain south of the Colorado River delta to high-relief cliffs south of Loreto where the Sierra de la Giganta touches the sea. This geomorphic variety is related to the tectonic forces shaping the Gulf of California, the arid climate of the region, and the presence of the Colorado River delta at the northern end. The eastern margin of the mountains that form the spine of the Baja California peninsula represents the edge of an extensional zone that was deformed during the opening of the Gulf of California. As the gulf opened, large tectonic blocks within this zone subsided into the newly forming ocean basin to varying levels. Some blocks slid deep into the new rift, while others became part of the shallowly submerged continental shelf. When the gulf's tectonic system became more transtensional in nature, some of these coastal and offshore blocks were elevated and offset in a variety of ways, giving us the coastal topography as well as the offshore shelf and islands we see today. The accommodation space available behind the beach and the width of the offshore shelf in front of the beach are controlling factors in the accumulation of coastal dunes. The shelf area upwind of the beach acts as both a sink for sediments delivered to the coast by local arroyos and, if the ecological conditions are favorable, the shelf becomes a place where biologically produced carbonate materials accumulate. Several examples from the peninsular coast of the Gulf of California illustrate the variety of ways that fetch, sediment availability, shelf area, and coastal topography affect the formation of dunes and dune fields.

El Gallo Dune Field

El Gallo dune field is located southeast of the Mulegé estuary behind a north-facing beach near the entrance to Bahía Concepción (fig. 10.1). This dune field is found within a low area of the coastline west of Cerro El Gallo (figs. 10.1B and 10.2). At least two episodes of dune development have occurred at El Gallo. An older dune field of unknown age is fully stabilized by vegetation, and a modern, partially vegetated dune field overlaps parts of the older dunes (Skudder et al. 2006). There are two areas of the modern dune field, a southern area that includes a sequence of three large transverse dunes, and a northern field that includes the hummocky foredune sitting behind the beach and a small transverse dune tens of meters behind the foredune (Skudder et al. 2006).

The foredune rises about 7 m high with small to large salt-tolerant bushes stabilizing areas of the front and crest. Between the bushes are troughs or blowouts

Figure 10.4. View to the north from the top of the pocket dune on Isla Carmen. Photo credit: Markes E. Johnson

Figure 10.5. View from the north of the northern region of the El Quelital ramp dune. Photo credit: David H. Backus

through which sand is transferred downwind to other parts of the field. Many of the channels established at the top of the foredune extend for tens of meters into the back-dune area parallel in direction to lines of bushes that stabilize the channel edges. These lines of vegetation are visible on satellite as well as other remotely sensed images and indicate the prevailing wind direction for the dune field. The orientation of the vegetation lines on the northern field indicates an average wind azimuth of about S15°E. The overall shape of both the older stabilized dune areas and the modern dune fields support this conclusion and are a good illustration of how the upwind and downwind areas of a dune area are affected by prevailing winds that blow at an oblique angle to the beach (Bauer and Davidson-Arnott 2002). In addition, the coastal mountains directly north of Mulegé and the promontory of Punta Chivato, which also lies to the north, shelter the western edge of El Gallo dunes and affect the morphology of the entire dune field. Along its eastern edge, sand blows well up the side of Cerro El Gallo more than 60 m above sea level.

Despite the fact that a broad marine shelf lies north of the beach at El Gallo, the carbonate content of the dunes is quite modest (5–15 percent by volume) compared to other dunes on the gulf coast of the Baja California peninsula (Skudder et al. 2006). It may be that fresh water or pollution from the Río Mulegé dampens marine productivity on the shelf in front of the beach at El Gallo.

Dunes at Punta Chivato

It is characteristic to find dunes perched on cliff tops along north-facing sections of the peninsular coast of the Gulf of California where sediment is available, but beach-level accommodation space is lacking (fig. 10.3). Russell and Johnson (2000) studied an example of this type of dune and the conditions under which it and other dunes formed at Punta Chivato. The promontory of Punta Chivato is located about 20 km north of Mulegé. Part of an uplifted tectonic block, the area is mapped as a series of andesite islands of Miocene age draped or covered by limestone and other sediments of Pliocene and Pleistocene age (Simian and Johnson 1997).

In contrast to the coast at El Gallo, steep cliffs back the beaches on the northern edge of the Punta Chivato promontory. Two narrow beaches along the northwestern edge of the promontory in low ground

between two paleoislands account for the only accommodation space available for dunes near sea level (fig. 10.1C). Low dunes (< 1 m high) are found behind the Ensenada El Muerto beaches. Most of the sand that arrives on the beach does not stay there, however, but is blown up and onto the cliffs behind the beach (Russell and Johnson 2000). With the exception of Isla San Marcos, which provides limited protection for the western edge of the Punta Chivato area, there is no impediment to the wind for hundreds of kilometers to the north. As was the case at Cerro El Gallo, the winter winds are able to transport sand up onto the flanks of the Punta Chivato promontory. Above the Ensenada El Muerto beach, sand is blown up to and above the 60-m mark, while on the marginally more sheltered eastern side of the promontory, sand has accumulated on a high terrace at the northeastern corner of the promontory, and blown over the 20- to 40-m high ridge behind the beach at Playa Cerotito (Russell and Johnson 2000). Several falling or slope dunes also are found in sheltered areas behind the ridge.

In contrast to the dunes at El Gallo, the composition of the dune-forming sediments at Punta Chivato is notable for the high percentage of carbonate content (50–75 percent; Russell and Johnson 2000). The high volume of coarse shell debris, particularly from bivalves, on the beach at Ensenada El Muerto implicates the existence of a substantial clam bank on the marine shelf to the north.

Pocket-Beach Dunes of Islas Carmen and Monserrat

The Baja California coast shows evidence of tectonic stresses related to the opening of the Gulf of California in many ways. Large-scale offset of tectonic blocks may or may not create coastal beach settings suitable for the creation of dunes like those at El Gallo and Punta Chivato. On a smaller scale, however, there are many places on the peninsular coast and gulf islands where pocket beaches have formed. Pocket beaches often develop where faults or fractures in the rock create a weakness that is widened into a narrow inlet by wind and wave energy.

Isla Carmen lies east of Loreto (fig. 10.1). One of the most unusual dunes we have encountered sits behind a pocket beach on the northern shore of the island (figs. 10.1 and 10.4). The beach and dune sit astride a north–south trending fault that cuts across the northeastern part of the island. The fault locally offsets Miocene andesite rocks and consolidated Pleistocene

Figure 10.6. Processed Enhanced Thematic Mapper image (bands 5,3,1) of the El Quelital ramp dune in Baja California and the Bahía Las Animas dune field to the north. Features associated with the sand ramp are a pluvial lake (PL), a falling dune (FD), and an area of the ramp that is falling (FR). Notice the comparative sizes of the sand ramp and the Bahía Las Animas coastal dune field. Image acquired from University of Maryland Global Land Coverage Facility. The original data set was collected by the NASA LANDSAT Program, LANDSAT ETM+ scene L7CPF20001001_20001231_08, L1G, USGS, Sioux Falls, November 13, 2000.

dune deposits. The inlet in front of the beach is long and narrow, with walls of white cross-bedded eolianite. Extending toward the south, the dune is about 185 m long and rises like a ramp to 17.5 m in height. The dune sand is 68 percent carbonate by volume and derived primarily from the skeletal debris of mollusks, particularly clams. Given the high carbonate content of this dune, a major clam bank probably exists on the marine shelf north of the island. Material eroded from the walls of the inlet also is an important source for carbonate material incorporated into the dune.

Located 45 km southeast of Loreto (fig. 10.1), the northern shore of Isla Monserrat features a 1-km-long beach with low dunes that rise behind to a height of 22 m above sea level. The beach and dunes are interrupted by Pliocene limestone that gives the effect of two adjacent pocket beaches. The dune sand here is 42 percent carbonate by volume and derived primarily from the skeletal debris of mollusks, particularly clams.

Coastal Ramp Dunes and Sheet Dunes

The least common but largest and most spectacular dune-related structures on the Baja California peninsula are sand ramps (fig. 10.5). Lancaster and Tchakerian (1996) were first to use the term "sand ramp" based on studies in the Mojave Desert of the southwestern United States. Sand ramps are layered deposits composed of eolian sand, talus from nearby mountain slopes, fluvial sediments, colluvium, and occasional paleosols (Tchakerian 1991; Lancaster and Tchakerian 1996). The diversity of sedimentary layers in a sand ramp reflects the variety of sedimentological processes found on the piedmont of the arid intermontane basins where they form. Lancaster and Tchakerian (1996) considered sand ramps to be part of a broader range of deposits formed through the interaction of sand-laden winds and topography. Deposits that include topographically controlled dunes can be dominated by eolian sands (> 90 percent; e.g.,

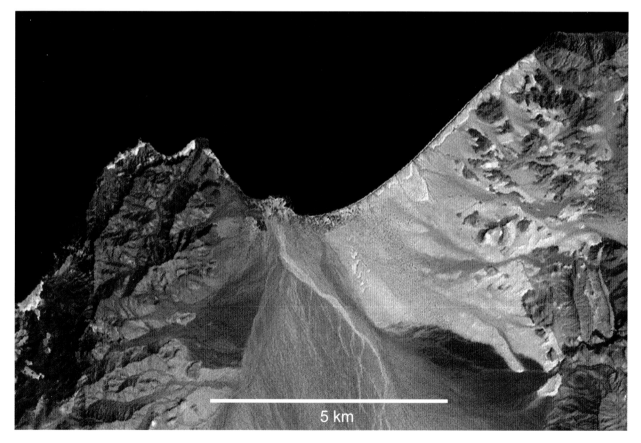

5 km

Figure 10.7. Processed Enhanced Thematic Mapper image (bands 7,4,1) of the relict ramp dune area on the northern side of Isla Tiburón. Image acquired from University of Maryland Global Land Coverage Facility. The original data set was collected by the NASA LANDSAT Program, LANDSAT ETM+ scene L7PCF20000719_10, L1G, USGS, Sioux Falls, September 3, 2003.

Figure 10.8. Processed Enhanced Thematic Mapper image (bands 5,3,1) showing the beach and dune ridge complex at Punta Arena, as well as the sand ramp (SR) deposits near Punta Colorada. Original data set was collected by the NASA LANDSAT Program, LANDSAT TM scene PO33R44_4T920604, Geocover, USGS, Sioux Falls, June 4, 1992. This image was resampled (Gram-Schmidt spectral sharpening) using band 8 from a second data set collected by NASA LANDSAT Program, LANDSAT ETM+ scene L7CPF19991001_19991123_12, L1G, USGS, Sioux Falls, October 14, 1999. Both images acquired from University of Maryland Global Land Coverage Facility.

climbing and falling dunes) or have eolian sand as a minor component (1- to 3-m-thick unit) of an alluvial fan sequence. Most of the sand ramps in the Mojave Dessert are relict deposits that accumulated during the Late Pleistocene and Early Holocene (Tchakerian 1991; Lancaster and Tchakerian 1996, 2003).

Although sand ramps were thought to be unique to the Mojave Desert, Backus et al. (2007) identified multiple sand-ramp deposits in the intermontane basins of northern Baja California, near San Felipe on the Gulf of California, on the northern coast of Isla Tiburón, and near Punta Colorada in Baja California Sur (fig. 10.1) based on LANDSAT and ASTER satellite data and images from the Google map service. Gastil et al. (1973) recognized some of

these sand ramps as dune areas in their geological reconnaissance mapping of the state of Baja California. We have also identified several broad, heavily vegetated coastal landforms that appear to be the near-shore remnants of coastal ramp dune deposits at La Bocana and Datilito in Baja California Sur (fig. 10.1).

All of the ramps from the interior mountain basins of Baja California appear to have accumulated under the same geomorphic conditions typical for sand ramps in the Mojave Desert. The ramps are all downwind from playas and sit within long, generally north–south trending basins that act as excellent wind corridors. The playas have very bright spectral signatures in satellite images due to evaporites left behind as former pluvial lakes dried out. We surmise that the same humid/arid cycles that affected the Great Basin of the southwestern United States during the Late Pleistocene also created these temporary pluvial lakes within intermontane basins of the northern Baja California peninsula. The lakes acted as sediment sinks when the climate was more humid and then became sediment sources for the sand ramps when conditions became more arid, as today (Tchakerian and Lancaster 2002). Deflation of alluvial fan deposits also would contribute sediment to the ramps (Lancaster and Tchakerian 1996).

Figure 10.6 is an Enhanced Thematic Mapper image showing a large sand ramp near El Quelital and its geomorphic relationship to the Gulf of California. The base of the sand ramp is about 18 km south of the beach at Bahía Las Animas. The dimensions of the ramp are approximately 9.5 km long from toe to southernmost end and 10.5 km wide. The ramp has blown over a spur of mountains along its western flank. A large falling dune can be recognized on the southern side of the spur, where the sand ramp approaches the ridge top on the northern side (fig. 10.6). East of the falling dune near the middle of the sand field, the ramp rises and then falls as it engulfs a lower portion of the ridge. This ramp appears to have been created primarily by northerly winds from the Gulf of California that were funneled down the Valle Las Animas from Bahía Las Animas. The bright spot on the northeastern edge of the ramp is the partial outline of a pluvial lake that is the probable source for the sediments incorporated into the ramp (fig. 10.6, PL). The southern side of the lake appears to be covered by an alluvial fan composed of sediments reworked from the ramp. The ramp rises approximately 300 m from its base near the lake to its southernmost edge. A visual estimate suggests to us that in places the thickness of the ramp is on the order of 75 m. Sand ramps from the Mojave Desert profiled by Lancaster and Tchakerian (1996) range from 5 to 70 m in thickness. Small patches of material that look similar to the sand ramp can be seen north of the lake and west of the ramp, suggesting that the ramp was more laterally extensive at one time. In fact, there are remnants of a falling ramp on the southern side of the mountain ridge currently abutted by the ramp, implying that the El Quelital structure once extended southward over the mountains and into the next valley.

Near the coast another area with a bright white spectral signature appears to be the remnants of a second pluvial lake or playa. At a much smaller scale than the sand ramp are coastal dunes that can be seen behind the barrier beaches at the mouths of the estuary and on the northern side of the promontory, which projects into Bahía Las Animas from the west (fig. 10.6).

Sand ramps near San Felipe and on the northern coast of Isla Tiburón appear to have the same morphology as intermontane ramps from the Mojave Desert region but a different source for their eolian sediments (fig. 10.7). We theorize that the ramps at San Felipe received material from the strip of tidal to subtidal sands that run parallel to the coast from north of San Felipe to the Colorado River delta (Thompson 1968). These shelf sands are likely to have been exposed during periods of low global sea level associated with glacial maxima during the Late Pleistocene and Early Holocene. Isla Tiburón sits in close proximity to the Sonoran coast on the gulf connected by a shallow marine shelf (Shepard 1950; Gastil and Krummenacher 1977). We infer that the shelf north of Isla Tiburón acted as the sediment source for the ramp under the same conditions proposed for the ramps at San Felipe. Also, in contrast to the ramps at San Felipe, the ramp sands at Isla Tiburón show an overall brighter spectral signature in satellite images that suggests a higher carbonate component in the dune sediments. The ramp at Isla Tiburón is highly dissected and more poorly preserved than the ramp at El Quelital or even the ramps at San Felipe. This difference may be related to the significantly greater amounts of rain that fall on the Sonoran side of the Gulf of California on a yearly basis (Hastings and Turner 1965; Douglas et al. 1993).

Not all coastal sand-ramp deposits we have observed sequester carbonate material, despite their proximity to a highly productive ocean. The multiple sand-ramp deposits near Punta Colorada in Baja California Sur are shown in figure 10.8. These spectrally dull-looking deposits lie within a low area stretching from the town of La Ribera to the southern edge of Punta Arena. The broad arroyo northwest of La Ribera is the ultimate source for most of the silica-rich sediment in the ramp deposits. The lack of carbonate sediments in the Punta Colorada sand ramps reflects the strong influence of the region's tectonic and erosional history.

Given that the sediment sources for the San Felipe, Isla Tiburón, and Punta Colorada eolian sand deposits would depend on changes in global sea level related to glacial cycles, it is likely that the pattern of accumulation would be similar but not necessarily in step with sedimentation patterns for the intermontane ramps of the northern Baja California region

Table 10.1. List of sand dunes on the Peninsular gulf coast of Baja California Sur with data on total surface area (in hectares), ratio of unvegetated to vegetated area, maximum dune height, estimated volume, and compass direction of dune migration. Numbers start with the south-most dune on the peninsula.

Name	Area (ha)	U/V ratio	Height (m)	Volume (m³)	Compass direction
1. Boca del Salado	0.58	3:1	0.1–2	894,000	180
2. Los Frailes (a)	0.01	1:0	2–4	30,000	180
3. Los Frailes (b)	0.02	1:0	2–4	60,000	180
4. Los Barracas	1.20	1:2.2	0.1–2	481,000	180
5. Punta Arena	9.78	1:2.9	0.10	489,000	160
6. Salinas P. Colorada	8.38	1:104	0.1–2	914,000	150
7. Laguna Salina (a)	0.29	1:6.3	0.1–2	67,000	180
8. Laguna Sallina (b)	0.07	1:2.5	0.1–2	26,000	170
9. La Capilla	0.86	1:5.6	0.1–4	339,500	175
10. Las Palmas	0.08	1:0	0.1–2	84,000	
11. Punta Pescadero	0.06	5:1	0.1–2	53,500	180
12. Salinas P. Arena (a)	3.81	1:3.6	0.1–2	1,170,000	155
13. Salinas P. Arena (b)	0.96	1:1.3	0.1–2	485,500	175
14. La Bocana	0.95	1:2.2	0.1–2	380,000	175
15. Punta Los Muertos	0.08	1:0	0.1–4	164,000	180
16. Boca del Rosario (a)	0.06	1:0	2–4	180,000	150
17. Boca del Rosario (b)	0.01	1:0	1–2	10,500	170
18. Ensenada El Coyote	0.09	1:0.8	0.1–2	104,000	175
19. Canal S. Lorenzo (a)	0.29	0:1	0.1–2	304,500	185
20. Canal S. Lorenzo (b)	2.21	0:1	0.1	221,000	185
21. Canal S. Lorenzo (c)	1.29	0:1	0.1	129,000	190
22. La Paz sand spit	15.03	1:4	0.1–2	1,200,000	190
23. Datilto (west)	0.63	1:25	0.1–2	234,000	190
24. San Evarista (a)	0.11	1:1.2	0.1–2	58,000	185
25. San Evarista (b)	0.14	1:1.8	0.1–2	61,500	175
26. No name	0.34	1:2.4	0.1–2	129,000	160
27. No name	0.04	1:10	0.1–2	8,000	170
28. Los Burros	0.13	1:1.6	0.1–2	60,000	180
29. Punta Malpaso	0.26	1:7.6	0.1–2	57,000	180
30. Punta El Cochi	0.19	1:5.3	0.1–2	47,500	185
31. Punta El Gato	0.08	1:7	0.1–2	17,500	185
32. Punta San Telmo	0.50	1:15.6	0.1–4	137,000	170
33. Punta Pasquei (a)	0.38	1:8.5	0.1–2	76,000	165

Table 10.1. *continued*

Name	Area (ha)	U/V ratio	Height (m)	Volume (m³)	Compass direction
34. Punta Pasquei (b)	0.40	1:4.7	0.1–2	106,000	165
35. Ensenada San Basilio	0.07	1:2.5	0.1–2	26,000	175
36. Ens. Puerto Almja	0.20	1:25	0.1–2	5,200	
37. Punta El Pulpito	0.08	1:7	0.1–2	18,000	175
38. Bahía San Nicolas (a)	1.26	1:6.7	7.0	626,417	170
39. Bahía San Nicolas (b)	0.41	1:9.3	2–4	161,000	170
40. Bahía Concepción	0.99	1:5.6	0.1–2	241,500	125
41. Playa Armenta	0.03	1:2	0.1–2	12,500	170
42. El Gallo (a)	1.24	1:2		6,550,000	165
43. El Gallo (b)	1.58	1:1.7		4,500,000	165
44. Punta Cacarizo	0.01	0:1	0.10	500	180
45. Playa Cerotito	0.07	0:1	0.1–5	178,000	175
46. Punta Chivato	0.01	1:0	0.1–2	?	
47. Ensenada El Muerto	0.53	1:1.9	0.1–2	224,000	170
48. Estero San Marcos	2.69	1:28.8	11.5–18	1,100,000	165
49. San Bruno	0.20	1:6.3	0.1–2	45,000	165
50. Estero Santa Ana	1.08	1:53	0.1–2	127,000	175
51. Ensenada la Trinidad	0.36	1:5	0.1–2	93,000	
52. Twenty-eighth parallel	0.41	1:7.2	0.1–2	88,500	175

Table 10.2. List of sand dunes on the Peninsular gulf coast of Baja California (northern state) with data on total surface area (in hectares), ratio of unvegetated to vegetated area, maximum dune height, estimated volume, and compass direction of dune migration. Numbers continue in sequence northward from dunes listed in Table 10.1.

Name	Area (ha)	U/V ratio	Height (m)	Volume (m³)	Compass direction
53. San Juan Bautista	?	0:1	0.10	166,000	175
54. Cabo San Miguel	?	0:1	0.10	37,000	165
55. San Francisquito (a)	0.06	0:1	0.10	6,000	155
56. San Francisquito (b)	0.13	1:3.6	0.1–2	44,500	155
57. San Francisquito (c)	0.05	1:1.5	0.1–2	24,000	
58. San Francisquito (d)	0.20	0.1	0.10	20,000	175
59. Cala Mujeres (a)	0.06	1:10	0.1–2	10,250	175
60. Cala Mujeres (b)	0.03	1:2	0.1–2	12,500	175
61. Punta Ballena	0.40	0:1	0.10	40,000	160
62. Marisma San Rafael	?	0:1	10.00	1,350,000	180
63. Bahia las Animas	0.18	1:3.5	0.1–2	56,000	165
64. Ens. El Quemado	?	0:1	0.10	30,000	
65. Bahía los Angeles	1.73	0:1	0.10	173,000	170
66. Ens. Alcatraz (a)	0.16	1:4.3	0.1–2	44,500	180
67. Ens. Alcatraz (b)	0.05	1:0.6	0.1–2	33,500	175
68. Bahía Calamajue	0.32	1:15	0.1–2	50,000	
69. Punta Final (a)	?	0:1	0.10	16,000	
70. Punta Final (b)	1.41	1:4.6	0.1–2	378,000	175
71. B. San Luís Gonzaga	?	0:1			170
72. Ens. San Francisquito	0.15	1:4	0.1–2	?	
73. Playa Bufeo	?	0:1	1.0	930,000	170
74. Bahía San Felipe	?	0:1	0.5	570,000	170

supplied primarily by pluvial lakes. Further study of Baja California's coastal sand ramps will be necessary to properly describe the relationship between their development and Pleistocene-Holocene glacio-eustatic sea-level changes.

Dune Ridge Sequence at Punta Arena

Development of several dune fields near the southern tip of the Baja California peninsula appears to be related to the erosion and redistribution of sediment along the coast by wave trains entering the gulf from the southeast, as well as waves generated by the northwesterly winds blowing down the length of the gulf. The best example of this phenomenon is the approximately 9-km² dune field at Punta Arena, where a series of shore-parallel dune ridges has prograded from west to east into the gulf (fig. 10.8). The source area for this dune field appears to be an older sheet dune or sheet deposit inshore of the dune field, as well as material from the marine shelf that lies upwind and to the north. Based on the topographical relationship between the dune field and the playa to the west, sediment for the dunes was probably supplied through nearby coastal estuaries until the prograding dune ridges cut off the estuary channels. Sediment

Table 10.3. List of sand dunes on islands in the Gulf of California with data on total surface area (in hectares), ratio of unvegetated to vegetated area, maximum dune height, estimated volume, and compass direction of dune migration. Numbers continue in sequence from dunes listed in Table 10.2.

Name	Area (ha)	U/V ratio	Height (m)	Volume (m³)	Compass direction
75. Isla Espíritu Santo (a)		0:1	0.50	295,000	185
76. Isla Espíritu Santo (b)					30
77. Isla Monserrat		0:1	0.50	30,000	170
78. Isla Carmen (a)		1:0	5.00	75,000	170
79. Isla Carmen (b)		1:0	18.00	172,800	175
80. Isla Coronados (a)	1.50	0:1	4.00	30,000	165
81. Isla Coronados (b)		1:0	2.00	50,000	
82. I. Angel Guarda (a)		1:0	1.00	15,000	165
83. I. Angel Guarda (b)		0:1	0.50	50,000	
84. Isla Granito		1:0	1.00	10,000	165

now appears to be supplied to the active portion of the field by the very prominent arroyo north of Punta Colorada and the redistribution of sediment along the seaward edge of the promontory.

The Punta Arena dune field is strikingly similar in morphology to the coastal dune sequence found at Magilligan Point in Northern Ireland, which also formed as the result of two intersecting wave fields (Carter and Wilson 1990). The beach and dune ridges at Magilligan Point are about 8 km at their maximum width and formed in several stages over approximately the last 5000 years. A future study that compares the temperate beach and dune sequence at Magilligan Point with the approximately 3-km wide, subtropical dune-ridge sequence at Punta Arena may provide additional insight into the origins of this type of coastal dune deposit.

Dune Field Volume and Composition

The results of a general survey of dune deposits on the peninsular coast of the Gulf of California are presented in tables 10.1 and 10.2. These may not include every dune deposit on the coast, as the search was limited by the resolution of the satellite images (15 m/pixel) and by our abilities to identify the variety of dune deposits found along the gulf margin. We have not seen every corner of the peninsular gulf coastline, but we have attempted to identify most of the major deposits to make a reasonable first-order estimate of the volumes of sand sequestered in coastal dune deposits. For expedience, we present our accounting of the dune fields found on the islands within the Gulf of California in table 10.3. Our survey includes both the less stabilized, active dune fields and the more heavily vegetated and older stabilized dune fields that are often but not always found associated with active dune deposits. In some cases, the younger deposits overlay older deposits to a considerable degree, so that the vegetation index used in tables 10.1–10.3 is not properly reflective of the true volume of the more heavily vegetated and stabilized deposits. We estimate on the order of 16.5 million m³ of sand is deposited in the active dune fields along the gulf coast of the Baja California peninsula, not including the coastal sand ramps. Another 8.5 million m³ of sand may be incorporated in the older stabilized coastal dune fields, excluding the coastal sand ramps. Data from table 10.3 on the gulf islands add another 1 million m³ of sand for both the younger, active and older, more stabilized dune fields. A grand total of 26 million m³ sand is calculated for the more common coastal dune fields found along the peninsular coast of the Gulf of California.

This total is significantly lower than the volumes of sand that are or were held historically in coastal or intermontane sand ramps. For example, the El Quelital sand ramp covers about 45 km² in area and holds three orders of magnitude more sand (940 million m³) than the average peninsular coastal dune field (0.5 million m³), almost two orders of magnitude more than the combined volume totals for the two dune fields at El Gallo (10.1 million m³), and an order of magnitude more than the combined total volume (tables 10.1–10.3) of all the other dune fields. Other coastal sand ramps that we have identified may have held comparable volumes in the past. At this point, we can only broadly compare surface coverage for those ramps we have seen first-hand.

Using the distribution of ramp remnants as a guide, it is estimated that the coastal sand ramp on Isla Tiburón covered at least 24 km², while the two ramps in the San Felipe area have a minimum combined coverage of about 43 km². By comparison, the relict ramp deposits near Punta Arena cover about 9 km² in area. Estimates of the carbon potentially held in these deposits currently are not possible because sand samples have not been studied. Previous work at El Gallo and San Nicolás by Skudder et al. (2006) and Punta Chivato by Russell and Johnson (2000) suggested the potential range in the carbonate content (15–85 percent by volume) of coastal sand dune deposits. Work of Skudder et al. (2006) also illustrated how a difference in the percentage of carbonate content affects the estimation of the amount of carbon held in a dune field.

At El Gallo, the total sand volume of the northern and southern dune fields (6.5 million m³) holds approximately 1.16 million metric tons of carbon, whereas the much smaller dune field at San Nicolás (350,000 m³) is estimated to hold a respectable 320,000 metric tons of carbon. It can be estimated that during the Pleistocene and Holocene, the coastal sand ramp at Isla Tiburón may have held four to five orders of magnitude more carbon than is currently estimated for the dune fields of Cerro El Gallo near Mulegé.

Dune Vegetation and the Coastal Wind Field

Because of their moisture-holding capabilities, dune sands are conducive to plant growth (Lancaster 1995). Dune vegetation is an important factor in the development of dune fields as a stabilizer of the dune surface and a modifier of dune morphology. Plant types found on dune fronts and in backdune environments can be divided into general groups. Ephemeral herbs have shallow root systems that benefit from heavy rainfall. Perennial plants that depend on the moisture stored in the dunes have single deep-penetrating taproots. There are also grasses that in some areas use

rhizomes to form a dense, dune-stabilizing mat or grow as clumps to minimize exposure and stave off burial by moving sand (Johnson 1977; Ash and Wasson 1983).

The flora of the Baja California peninsula has been described in a compendium by Wiggins (1980). Major studies on dune vegetation within the region were done by Johnson (1977, 1982), who surveyed the Pacific coast of the peninsula and western side of the Gulf of California as far north as Bahía La Paz. Johnson (1977) surveyed both foredune and backdune environments of the major dune fields she encountered. While the perennial herb *Abronia maritima* was ubiquitous as a stabilizer of the foredune along the Pacific coast of the Baja peninsula, she noted several major changes in the other species associated with foredunes. Breaks in vegetation zones occurred around 30°N, at 28°N near the border between the state of Baja California and Baja California Sur, and at 24°N, putting the cape region in a separate zone (Johnson 1977). South of 24°N, the grasses *Jouvea pilosa* and *Sporobolus virginicus* become the dominant stabilizer of foredunes.

Our particular interest in vegetation is because of its usefulness in illustrating coastal wind patterns. Johnson (1977) noted that sand transported to the back of the dune is channeled between the vegetation that stabilizes the foredune, as affirmed in our description of El Gallo dune field. As a dune field evolves, a linear pattern develops in the vegetation of the backdune area that parallels the dominant wind direction for the dune field. This linear pattern is easily seen in remotely sensed images with sufficient resolution. A more recent study of the coastal dunes at Punta Chivato by Russell and Johnson (2000) used the orientation of leatherplant (*Jetropha cuneata*) and copal trees (*Bursea hinsinia*) deformed by windborne sand and salt aerosols to map the local wind field.

An azimuth for each dune field is recorded in tables 10.1–10.3, showing the median direction that sand travels across the area. The overall shape of the dune field was also used to corroborate our findings. A summation of wind-azimuth data is provided in figure 10.1D. The mean value for our data is S8°E, indicating that the dominant wind source for the dune systems of the peninsular coast of the Gulf of California is from an average wind direction of N8°W. Russell and Johnson (2000) report the mean wind direction for Punta Chivato as from azimuth N2°E, while the average wind direction for the entire Gulf of California was reported as azimuth N24.5°W by Parés-Sierra et al. (2003). Variations in the local wind-field data represented in tables 10.1–10.3 are due to the modifying influences of local topography on the overall Gulf of California wind field.

There are two locations where the local wind field deviates significantly from the bulk of the data. The first is a dune field that sits behind a ridge that runs southwest–northeast on Isla Espíritu Santo on the eastern side of Bahía La Paz. The configuration of this dune field is probably the result of the westerly wind known to cool the city of La Paz in the afternoon during the summer months (Roberts 1989). The second dune field lies at the base of Bahía Concepción, where the mountains of the Concepción Peninsula funnel gulf winds down the 40-km-long, northwest–southeast oriented embayment. On the whole, the topography of the Baja California coast hems the edge of the Gulf of California wind field and deflects the northwesterly winter winds so that they becomes more southerly in flow.

11 Beach Deflation and Accrual of Pliocene-Pleistocene Coastal Dunes of the Gulf of California Region

Markes E. Johnson and David H. Backus

Nothing would seem more ephemeral on a geological time scale than the seasonal winds that blow out of the north across the Gulf of California during the long winter months from November to April. As reported in chapter 10 on coastal sand dunes, winter winds persist for days at a time, commonly gusting to 8 m/sec or higher under sunny skies that otherwise betray no hint of storm conditions. Yet, the tremendous fetch over which the winds follow the axis of the gulf are capable of agitating huge swells more characteristic of storms at sea. Sand migration tracked on 84 coastal dunes (see tables 10.1 and 10.2) scattered over the entire gulf coast and associated islands confirms an overall southerly direction of movement, which means that north-facing beaches absorb the full force of seasonal wind and wave activity. During placid days that typically intervene to make the Gulf of California such an agreeable place to visit in wintertime, the many small to medium-sized dunes that exist from place to place remind us of the strong *norte* winds. Some dunes accumulate in dead-air space on the leeward side of rock ridges. Thus, the norte winds are strong enough to carry sand grains off the beach and inland over topographic barriers.

As dune sand becomes cemented to form deposits of solid rock and undergoes further diagenetic changes, processes are set in play that require geological time measured in thousands to millions of years. The number of localities where fossilized dune deposits are known to exist around the gulf is small in comparison with the number of dune fields found in present-day coastal areas. The remains of sheet dunes younger than 125 Ka are commonly associated with many modern dunes in the Gulf of California, as discussed in chapter 10. Walker and Thompson (1968) documented the stratigraphy of Pleistocene dunes from the near Recent (less than 42 Ka) in the upper Gulf of California. Dune rocks from Baja California Sur are more strongly lithified and significantly older than the semiconsolidated dune deposits from the near Recent. That former dunes were preserved in any configuration at all, however, testifies to the long record of norte winds in the gulf region.

Like the region's contemporary dunes, the fossilized dunes also show concentrations of sand derived from the skeletal debris of marine life with calcium carbonate shells, tests, or other hard parts. Yet again, this aspect demonstrates the linkage between the gulf's present and past fecundity. Through summary of earlier studies on dune rocks in Baja California Sur and generation of new data on the composition and volume of those deposits, in this chapter we illustrate something of the geological history that preceded the allocation of coastal dunes in today's gulf.

Physical and Biological Evidence for Dune Rocks

Recognition of former sand dunes in the rock record is based on criteria using both physical and biological evidence. In its dispersal, dune sand is thoroughly sorted by the wind, which means that sand grains are relatively uniform in size at any given place within a dune. Cross bedding of well-sorted sand is a diagnostic feature used to identify fossil dunes. Cross bedding results from the way in which dune sand migrates downwind. Most dunes typically show a longitudinal dichotomy with a low-angle windward side called the stoss slope and a high-angle slip face called the lee slope. The ramp angle on a stoss slope might be 10° to 15°, whereas the lee slope generally is maintained near the angle of repose close to 33° (Kocurek 1996). Sand moves up the gentle stoss slope by saltation until it reaches the dune crest, where it collects near the top of the lee slope. The upper lee slope is destabilized when too much sand accumulates and gravity pulls material down the slip face in a small-scale avalanche.

Layers accrete on both parts of the dune, but as the dune migrates downwind the low-angle bedding from the stoss slope overrides the high-angle bedding from the lee slope. In cross section, this arrangement forms a sharp line of truncation between strata with two very different dip angles.

Dune fields develop as morphological types that include families of crescentic, linear, star, compound, and complex dunes (Kocurek 1996). Compound and complex dunes entail a large geographic footprint less likely to achieve preservation, whereas the smaller varieties of crescentic and linear dunes are more prone to burial and potential preservation in the rock record. Thus, cross bedding in a rock formation may be linked on a larger scale to discrete rock bodies that conform to a particular style of dune morphology.

Although sand from an ancient dune may be composed of carbonate grains derived from marine organisms, accumulation is a land-based process. The fossils of whole marine shells are entirely lacking or very rare. Birds are known to carry shellfish inland, and the authors have observed seagulls around the dune fields south of Mulegé so engaged. The odd shell found in dune rock would be unexpected, but not improbable. Other sorts of fossils commonly occur in dune rocks, but they are indicative of a purely terrestrial environment. Invertebrate fossils may include pulmonate (air-breathing) gastropods, which tend to be small with thin shells. Root casts from the various plants either buried by the advancing dune or that had colonized the dune surface also are incorporated in dune rock.

Identifying Dune Rocks

The LANDSAT and Advanced Spaceborne Thermal Emission and Reflection Radiometer (ASTER) satellite images that proved to be an asset in the collection of primary data on present-day coastal dunes in the Gulf of California were less useful in the systematic search for dune rocks through the same region. Band combinations that aid in spotting contemporary dunes enriched to some degree in carbonate materials are supplemented by patterns typical for dune vegetation. Existing band combinations cannot distinguish between limestone associated with dune rock and marine limestone. However, satellite images that cover places where dune rocks are known to be extensive help to determine the surface area of those deposits and their remoteness from potential sources where

beach deflation originally occurred. Likewise, satellite images proved accurate in detecting small outliers separated by geographic barriers from the main body of dune rock.

As one of the goals of this study is to rate the enrichment of dune rocks in carbonate sand, it was critical to visit locations where measurements could be made to help calculate the overall volume of fossil dunes. A related goal is to determine the bulk population of whole mollusk shells or other organic entities required to form a given body of dune rock. The organisms contributing shell material were identified through the analysis of petrographic thin sections cut from dune-rock samples. The methodology for these calculations was developed by Skudder et al. (2006) using the common bivalve *Megapitaria squalida* as a model for sequestration of carbonate shell material in present-day coastal dunes. Sewell et al. (2007) formulated a similar procedure for the study of dunes containing rhodolith material. Computations offered here are the first of their kind related to input of whole organisms required to make dune-rock formations at specific localities in the Gulf of California.

Pliocene-Pleistocene Beach and Dune Rocks at Punta Chivato
Carbonate Beaches as Sediment Source

Near Mulegé (fig. 11.1A, PC), the Punta Chivato promontory entails rugged topography eroded in andesite bedrock covering about 25 km^2 (fig. 11.1B). Russell and Johnson (2000) surveyed this area for the distribution of beach sandstone and dune rocks. Deposits with high concentrations of carbonate sand were found to occupy Pleistocene terraces at three localities in the vicinity of Punta Chivato (fig. 11.1B). The sandstone at locality 1 south of the light tower at Punta Chivato sits on terrace conglomerate composed of andesite cobbles with integrated marine fossils 11 m above present sea level. With an overall carbonate concentration of 86 percent dominated by crushed bivalve debris, distinct layers of the coarse "sandstone" dip gently 4° seaward. Although a small, active sand dune is situated nearby, the carbonate rocks at this locality are regarded as a Late Pleistocene beach deposit that formed during the last interglacial epoch about 125 Ka (Russell and Johnson 2000). A similar deposit occurs at nearly the same elevation west of the Punta Chivato light tower at locality 2, but occupies a depression eroded in Pliocene marine limestone (fig. 11.1B). At locality 3 farther to the west, another carbonate-sand deposit sits higher against the hillside 27 m above present sea level. It represents a somewhat older Pleistocene beach that experienced more coastal uplift. All three localities show distinctive weathering called pinnacle karst: small limestone towers individually about 50 to 70 cm in height that formed due to dissolution after the beach deposits were cemented and uplifted. Prior to cementation, these beaches had the potential to feed coastal sand dunes.

Canyon Dunes

The rocky coastline that stretches 4 km west from Punta Chivato faces the Ensenada El Muerto, a great expanse of open water obstructed only by Isla Tortuga 40 km directly north. Four sizable canyons cut through 80 m of andesite basement rocks on the northern side of the Punta Chivato promontory, but only the westernmost canyon contains dune-rock deposits (fig. 11.1B). The norte winds that cross the 100-m-wide pocket beach at the mouth of the canyon deflate the beach and transfer sand inland. Near the mouth of the valley (fig. 11.1C, locality 1), dune rocks with a thickness of 3 m retain tabular cross-beds that dip 32° inland on leeward slopes. These rocks are incised by the lateral equivalent of the same 11-m terrace found south of Punta Chivato (fig. 11.1B). Likewise, the 11-m terrace at the valley entrance includes marine fossils associated with conglomerate. Here, however, the relationship shows that dune rocks already existed when the sea level rose high enough to notch the marine terrace. Thus, the dune rock must be older than 125 Ka.

The valley interior shows no other terraces like the 27-m terrace on the outer coast (fig. 11.1B, locality 3). An equivalent marine terrace should have developed in the outer part of the valley, if open when the 27-m terrace formed nearby. Lack of any higher terraces within the valley suggests that it was fully occupied by dune rocks at the time. This interpretation is supported by the fact that massive dune rocks reach an elevation approaching 60 m above present sea level on the western side of the valley (fig. 11.1C). The present arroyo cut through a more extensive deposit of dune rocks, exhuming the valley in many places to its original andesite floor and walls. Based on the record of global fluctuations compiled by Haq et al. (1987) and Miller et al. (2005), Pliocene sea level stood about 50 m below the present level at approximately 2.4 and

3 Ma (see fig. 3.2). At about 1.5 Ma during the Early Pleistocene, sea level may have been 100 m lower than today. The deep canyons on the northern side of the Punta Chivato promontory developed during the Late Pliocene to Early Pleistocene and acquired dune deposits during those intervals when sea level was the same as or lower than now. Beaches along the front of the Punta Chivato promontory would have been more extensive during periods of lower sea level.

A good example of a small fossil dune is preserved 400 m inland and 40 m above present sea level (fig. 11.1C, locality 2). Elsewhere, bedding on leeward slopes exhibits dips that point mainly southward. At this locality, however, what were linear dunes oriented with the narrow valley changed course to follow a more southwesterly direction where the valley widens out. Dune rocks with a leeward slope of 28° and an azimuth of S50°W are well exposed in profile over a horizontal distance of about 18 m (fig. 11.2). In a study of krummholz patterns (wind-sculpted bushes and small trees) on the Punta Chivato promontory, Russell and Johnson (2000) found that subtle landscape variations caused eddies that deviate from the dominant wind flow. At this particular location, it appears that migration of dune sand veered from a south to a southwest track in response to valley geometry. Locally, the dune rock is perforated by many vertical hollows with a diameter of 5–10 cm that resemble the shape and texture of small tree trunks.

Assessment of Organic Content in Dune Rocks

Based on a petrographic thin section from a dune-rock sample collected near the bottom of the fossil dune in figure 3.2, bioclasts that can be traced to particular organic sources for calcium carbonate account for 56 percent of counted grains (Russell and Johnson 2000). The remaining 44 percent are dominated by andesite derivatives. Within the carbonate fraction, most bioclasts come from bivalve shells (36 percent), but some also derive from coralline red algae (9 percent). Compositional analysis is complicated by the fact that a significant number of bioclasts were removed through dissolution or were replaced by another form of calcium carbonate through diagenesis. In outline, the ghost grains represented by vacuities typically resemble those expected from mollusk grains. As the ghost grains were not counted in the tally for bioclasts, the 4:1 ratio of bivalve to algal materials is undervalued with respect to mollusks.

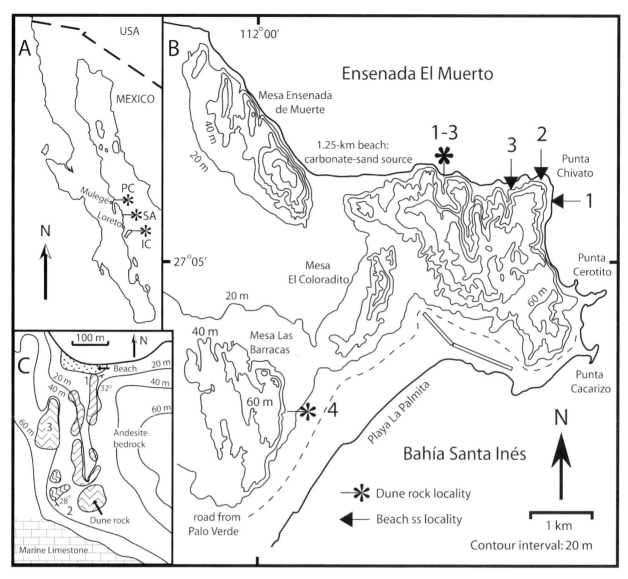

Figure 11.1. (A) Map including the Baja California peninsula with asterisks marking the location of Punta Chivato (PC), Punta San Antonio (PA), and northern Isla Carmen (IC); (B) complete topographic coverage for the Punta Chivato region with numbered localities for Pleistocene-Pliocene beach sand and dune rocks; and (C) enlarged view of valley with dune rocks on the northern side of the Punta Chivato promontory.

Assuming that the 4:1 ratio of bivalve to algal materials from the thin-section census is a conservative estimate and that the sample is reasonably representative of the larger dune-rock deposit from which it comes, it is instructive to further quantify the population of marine invertebrates and plants required to contribute to the calcium carbonate enrichment of the small dune from the inner part of the Pliocene-Pleistocene valley. The first step is to determine the volume of the fossil dune, only part of which can be seen in figure 3.2. The fossil dune has an outcrop exposure with a maximum thickness of 3.75 m and a surface area 50 m long by 19 m wide. A rectangular box with these dimensions holds 3,562 m³. The bulk contribution of mollusk-shell material to this volume at 36 percent purity amounts to 1,282 m³. Following the experimental work of Skudder et al. (2006) on mechanical reduction of bivalve shells, it takes a population of 15,650 mature *M. squalida* bivalves to produce 1 m³ of dune sand after the shells are disarticulated and crushed to a mixture of fine-to-coarse carbonate sand. Known as the chocolate shell in Mexico due to its coloration, this species is abundant in the Gulf of California today and is well represented also

Figure 11.2. Looking south at a geological section through dune rocks that show leeward beds (25–45 cm in thickness) with a dip angle of 28° on the leeward slip face. People for scale spread out over a distance of 18 m. Photo credit: David H. Backus

as a Pliocene fossil (see chap. 9). If the species were exclusively responsible for the carbonate material derived from mollusk shells in this body of dune rock (surely not), it would require a population in excess of 200 million clams.

Along the same experimental lines, Sewell et al. (2007) determined that it took 16,265 rhodoliths (spherical growths of coralline red algae) with an average diameter of 5 cm to make 1 m³ when reduced to sand-size fragments. The bulk contribution of crushed rhodoliths to the same fossil dune using the 9 percent figure would amount to 320 m³. This volume requires the equivalent of more than 5.2 million whole rhodoliths with a 5-cm diameter to be recycled as dune sand. Naturally, rhodoliths come in sizes both smaller and larger than the models employed in the experimental trials, but the raw numbers give an impression of the vast numbers of organisms that donate calcium carbonate to coastal dunes.

The massive block of dune rock on the western side of the valley (fig. 11.1C, locality 3) has walls that rise steeply above the 40-m contour line. As now exposed,

the block's surface area measures 80 m parallel to the valley wall by 60 m wide. Talus at the base of the cliff makes it impossible to judge the true thickness of the block, but it has an 8-m drop off. Using these conservative dimensions, the block has a minimum volume of 38,400 m³. If the dune rock from this block were 36 percent enriched by the shells of crushed *M. squalida*, the population needed to produce that portion of the former dune would amount to something in excess of 6 billion clams. Likewise, if the block were 9 percent enriched by the crushed debris of rhodoliths, that part of the former dune would entail the contribution of 56 million whole rhodoliths with an average diameter of 5 cm.

A loop in the 40-m contour line around the northern end of the block shows the extent to which the contact between dune rock and the valley's original andesite wall acts as a zone of weakness for accelerated erosion. Based on the pattern of the contour line and the open space adjacent to the north (fig. 11.1C, locality 3), it can be conservatively estimated that half the dune rock from the western wall of the valley has already been removed by erosion. Thus, the missing part doubles the estimate for this segment of the valley, which would entail sequestration of calcium carbonate from 12 billion clams and 112 million rhodoliths. The numbers become astronomical when it is considered that the greater part of the valley probably was packed with dune rock.

The most likely scenario for the development of these deposits is that changing sea level and varying angles in stream profile during uplift of the Punta Chivato promontory influenced multiple stages of dune development within the valley. It is likely that the dune rocks along the arroyo and the small dune near the rear of the valley (fig. 11.1C, from localities 1 to 2) represent a later stage of dune formation following exhumation that left some dune rocks in place at a higher elevation along the sides of the valley (fig. 11.1C, locality 3).

Dune Rocks Against Mesa Las Barracas Escarpment

A faulted succession of Pliocene strata that tilt gently to the southwest goes by the name Mesa Las Barracas in the southern Punta Chivato area (fig. 11.1B). The structure's master escarpment is an uneven cliff line up to 30 m high that extends for more than 2 km along the eastern side of the mesa. Resistant limestone bearing abundant marine fossils and softer limestone beds with extensive trace fossils are well exposed in the cliff face. All fossils date from the Late Pliocene. What on first impression appears to be a fault line bringing a radically different set of strata into contact with Mesa Las Barracas is actually a contact with Pleistocene dune rock that accrued against the escarpment (Johnson 2002a). Because the dune rocks are composed of softer limestone than the older Pliocene cliffs, erosion by wind and rain carved out a narrow defile between the two contrasting bodies of rock. It is possible to stand at the bottom of this trough (fig. 11.1B, locality 4) and look north between Mesa El Coloradito on the east and Mesa Ensenada de Muerte on the west to see gulf waters on the horizon. The beach at Ensenada El Muerto with its enormous store of coarse carbonate sand is 4 km away.

The dune rocks at locality 4 are not extensive in surface area, measuring only 15 × 25 m parallel to the adjacent escarpment. On a satellite image, the spot is a tiny cluster of pixels difficult to differentiate against

Figure 11.3. Looking north on a depression eroded in granodiorite rocks filled by dune rocks at Punta San Antonio. Cliff face on the dune rocks is about 4.5 m high. Photo credit: Markes E. Johnson

the larger patch representing the marine limestone of Mesa Las Barracas. Exposed thickness amounts to only 3 m, making the deposit insignificant in terms of volume. Besides the fact that the deposit is easily identified as dune rock on the basis of its uniform layers of well-sorted carbonate sandstone with an inclination of 30°, two other aspects are noteworthy. The dip angle is typical for the leeward slope of an advancing dune and the beds point southward. Given this drift direction, the more intriguing aspect of the locality is that the only possible source for the carbonate sand was a Pleistocene beach at Ensenada El Muerto 4 km to the north. Not only was the wind strong enough to move sand across that distance, but also up grade through an elevation of 40 m. An active dune of the same sort exists nearby with much the same dimensions and drift direction (Johnson 2002a, fig. 24).

Pleistocene Dune Rocks at Punta San Antonio

Less than 1 km² in area, Punta San Antonio protrudes into the Gulf of California roughly midway between Mulegé and Loreto (fig. 11.1A). Basement rocks composed of granodiorite dominate the point and form low sea cliffs rising from 10 to 15 m in height. A small body of dune rock is preserved on the southern side of the point, filling a gulley-like depression on the flanks of a pluton close to sea level (fig. 11.3). High content in calcium carbonate material is revealed by the incipient development of pinnacle karst on the eroded face of the dune rocks. In cross section, the fossil dune has a maximum height of 4.5 m and a maximum width of 16 m at the top of the coastal cliff line. The dip angle from bedding exposed in the transverse face of the dune rock cannot be determined, as the beds follow a scoop-like profile faithful to the contours of the original gulley. These dune rocks are interpreted as a falling dune that accumulated on the leeward side of Punta San Antonio in dead-air space. The carbonate sand that formed the dune would have come from a beach on the northern to northeastern side of Punta San Antonio, about 300 m away. To reach the shelter of the gulley, however, the sand would have migrated up and over the top of the granodiorite peninsula.

Pleistocene Dune Rocks on Isla Carmen

Anderson (1950, p. 20) recognized extensive deposits of Pleistocene dune rock on the northern end of Isla Carmen, which he described as "nonfossiliferous cross-bedded calcareous sandstone." Petrographic thin sections showed the dune rock to be composed of "rounded to subrounded shell fragments," but Anderson (1950) made no attempt to assess the level of carbonate enrichment. A false-color satellite image modified to characterize the local distribution of dune rocks is centered on the northern end of Isla Carmen (fig. 11.4). In this view, two white patches with vegetated lines in pale-green to purple colors are situated along Bahía Oto in the northwestern corner of the island and near Puerto de la Lancha to the northeast. Valleys with a north–south orientation, also containing active dunes, contribute to the shape of the dune-rock patches. Some valleys hold dunes that extend farther inland than others. Dune rock underlies much of the two areas, separated from one another by andesite ridges that appear orange to brown in the satellite image.

Puerto de la Lancha Area

Dune rocks cover approximately 2.3 km² (or 230 ha) immediately east of Puerto de la Lancha. Figure 11.5A shows a cross section through dune rocks exposed in a cliff along a narrow marine inlet. The dune rocks in this view preserve the leeward face of two distinct dune cycles, both with layers that dip 30° in the same direction to the south. High-angle cross bedding is seen in the lower, far-right area of the photograph. A clear line of truncation marks the passage of one dune cycle with its stoss slope overriding part of the landward lee slope. The height of this particular sea cliff is only 3.5 m. Fossil land snails (pulmonate gastropods) are known from the same general vicinity above the sea cliffs. Anderson (1950, p. 20) was the first to mention these Pleistocene fossils, which he referred to the genus *Bulimulus*. A more diverse fauna, including *Camaena* sp. also occurs in the Pleistocene here (fig. 11.5B). The marine inlet at this locality terminates at a 60-m-wide pocket beach behind which is an active dune described in chapter 10 (fig. 10.4).

Vertical sea cliffs that rise 70 m in height closer to Puerto de la Lancha record the complex interaction of as many as five individual dunes that followed one after another, as well as the relationship of the dunes to the underlying andesite basement rocks (fig. 11.6). Stretching for more than 300 m along the waterfront, the cross section from the sea cliffs shows the complete outline of a bow-shaped unconformity eroded by processes that deposited dune 2 atop dune 1. The

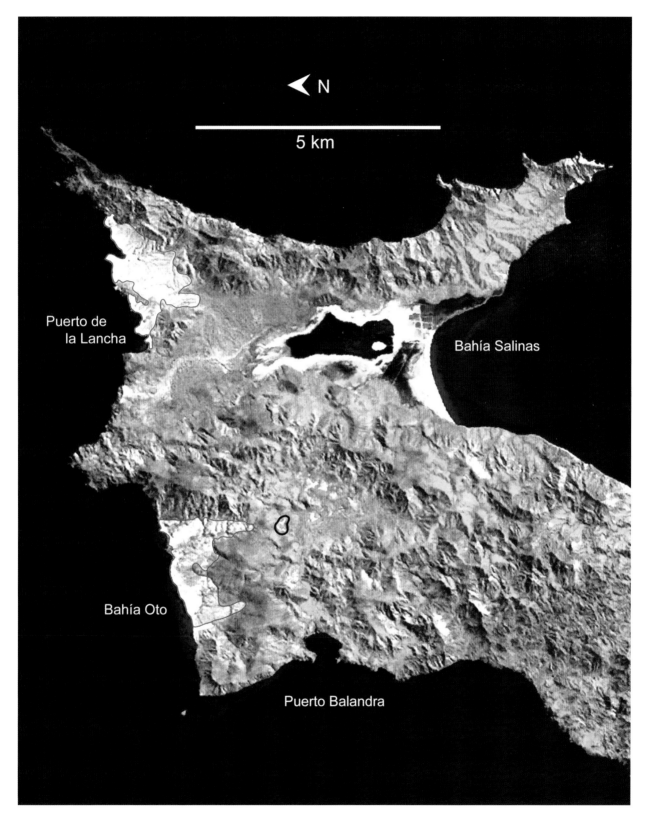

Figure 11.4. False-color satellite image for the northern part of Isla Carmen in the lower Gulf of California showing distribution of present-day sand dunes and underlying Pleistocene dune rock. Note location of major dune-rock deposits (outlined in red) to the northwest and northeast, which show as beige patches. The inland dune circled in black is shown in figure 11.7. Enhanced Thematic Mapper image acquired from the University of Maryland Global Land Coverage Facility. Original data set was collected by NASA LANDSAT Program, LANDSAT ETM+ scene L7CPF20011001_20011231_05, L1G, Sioux Falls, October 17, 2001.

Figure 11.5. (A) Northeastern Isla Carmen near Puerto de la Lancha looking west on sea cliffs 3.5 m in height that feature Pleistocene dune rock; note cross-bedded layers from separate dune generations (lower right and upper left) both with dip angles of 30° on the leeward slip face; and (B) apical view of Pleistocene pulmonate gastropod (*Camaena* sp.) from same general vicinity; shell diameter is 2 cm. Photo credits: David H. Backus

underlying dune rock is bifurcated by a high-angle reverse fault that brought andesite basement rocks closer to the surface on the western side of the fault, revealing an older unconformity with more localized topographic relief. Faulting probably controlled the position of the subsequent valley prior to deposition of dune 2. It appears that the calcareous sandstone in dune 2 accumulated to a minimum thickness of 36 m through the accretion of sand in subhorizontal beds on the low-angle stoss slope of a linear dune with a north–south orientation.

The cross-sectional area of dune 2 (fig. 11.6) amounts to approximately 4800 m². There is no way to appraise to what extent coastal erosion reduced the former valley or the size of the linear dune it contained. However, the satellite image in figure 11.4 shows that dune rocks near Puerto de la Lancha penetrate no less than 1 km inland to the south. Using this conservative estimate for the length of the sand body contained in dune 2, its present volume is calculated as 4.8 billion m³. If the Pleistocene dune rock represented by this single feature holds about the same concentration of organic detritus as the nearby active dune at 68 percent (see fig. 10.4), then dune 2 may contain 3.25 billion m³ of carbonate sand. Using the *M. squalida* bivalve model of Skudder et al. (2006) that requires a population of 15,650 mature bivalves to produce 1 m³ of dune sand after mechanical reduction,

more than 50 billion whole shells would be needed to enrich this single fossil dune. If all the superimposed fossil dunes spread out over 230 ha are speculated to have a composite average thickness of only 25 m, then the total volume of Pleistocene dune rock on the northeast corner of Isla Carmen would amount to a volume of 57.5 billion m³. At any level of carbonate concentration above 50 percent, the numbers of shelled mollusks necessary to provide that content is gargantuan. Of course, accretion of the fossil dune fields in this area took place over a long time, meaning that organic input was not limited to any one infusion but was more continuous in nature. Anderson (1950, p. 20) suggested that the older dune rocks near Puerto de la Lancha are from the Upper Pliocene.

Bahía Oto Area

Based on the satellite image in figure 11.4, dune rocks and associated active dunes cover an area of 2.15 km² (or 215 ha) directly south of Bahía Oto in the northwestern corner of Isla Carmen. During low tide, it is possible to access sea cliffs capped by dune rocks at the far western side of the bay. A limestone matrix (in which Pleistocene gastropods such as *Turbo fluctuosus* and other bivalves occur) holds together conglomerate composed of andesite cobbles and boulders up to 75 cm in diameter. With a maximum thickness of 2 m, the conglomerate sits directly on Miocene andesite from which the andesite boulders originated through vigorous coastal erosion. Immediately above the conglomerate is a bed of pink limestone 1.5 m thick composed of coarse sand-size particles derived from abraded mollusk shells. This layer is interpreted as beach sand that washed onto the conglomerate under strong wave assault. Overlying the beach sand in the cliff section is a massive, 4-m-thick deposit of beige dune rock extensively riddled by dissolved cavities due to karst erosion. Hand-lens examination of the dune rock shows it is composed of small grains of andesite and shell detritus less than 0.5 mm in diameter. Because the dune rocks are stratigraphically above the conglomerate with Pleistocene marine fossils, it means they cannot be older than the Pleistocene in age, as is true for all other dune rocks exposed at higher elevations within the Oto basin.

Dune rocks extend inland to the south about 800 m from this coastal section, where gently dipping beds with a minimum thickness of 2 m are exposed in profile above the talus line. Here, the dune rocks abut against andesite hills at the rear of the basin. Dune

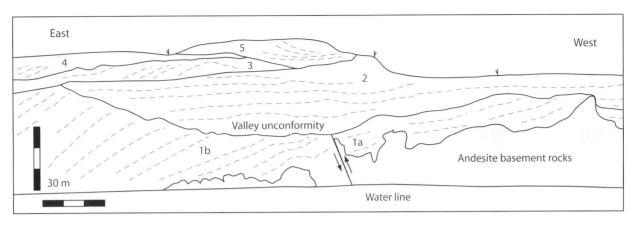

Figure 11.6. Geological cross section showing a succession of fossil dunes exposed in sea cliffs near Puerto de la Lancha on the northern coast of Isla Carmen. The interpretation was drawn from composite photographs looking due south from a seaward position. There is no vertical exaggeration in this overview, and the height of the sea cliffs was estimated on the basis of 3-m-tall cardón cacti growing at the edge of the plateau.

Figure 11.7. Karst-eroded dune rocks from the northwestern interior of Isla Carmen inland from Bahía Oto. A 200-m-high ridge separates this outcrop from the nearest beach source approximately 0.5 km away. Photo credit: David H. Backus

rocks on the far western side of the Oto basin abut against andesite hills 50 m above present sea level, but lateral equivalents appear not to be represented elsewhere through the basin. Due to the slightly smaller size of the Oto basin and its thinner deposits, the volume of dune rocks is still substantial but less than for the Puerto de la Lancha area.

Following the fault valley that runs inland east of Puerto Balandra (fig. 11.4), it is possible to reach an area isolated from the main body of dune-rock deposits on Bahía Oto by a range of coalescing andesite domes with elevations peaking about 245 m above sea level. The view in figure 11.7 looks to the north, with one of the red andesite domes in the background and a body of grayish dune rock in the middle distance. The maximum thickness of the dune rock at this locality is only 16 m, but it appears to be much greater due to the extensive talus pile below the outcrop on the side of the hill. The east–west length of the fossil dune field, however, is about 450 m. Circled by an identifying ring on the satellite image (fig. 11.4), the thin line of pixels representing this outlying dune-rock deposit can be seen to occupy a spot directly in line with dune rocks that penetrate farther south from the shores of Bahía Oto than in any other part of the basin on the opposite side of the intervening ridge. To reach this spot on the island's interior, sand was carried off the beach 2.5 km away. More surprising, the sand in this outlying Pleistocene dune was lifted over a substantial barrier at least 200 m in elevation. Thus, the eolian deposit represented in figure 11.7 can be classified as a falling dune that accrued in dead-air space on the leeward side of the andesite barrier. No other locality we are aware of attests so clearly to the dynamism of the norte winds in former Pliocene-Pleistocene times.

Summary of Major Relationships

Along the gulf coast of Baja California and its associated islands, fossil dunes more than 40 Ka in age are known only from the lower Gulf of California.

The small size of dune-rock deposits on north-facing coasts or even smaller outlying spots in leeward settings makes them difficult to locate using satellite reconnaissance techniques. Examples fall into two categories. Fossil linear dunes are constrained by narrow valleys that open northward onto beaches that supplied calcareous source materials. These valleys generally resulted from tilted tectonic blocks that promoted differential erosion in a north–south orientation. The longest such valley is 2 km in length off Bahía Oto on Isla Carmen. High-angle bedding with a dip close to 30° on the slip face is the most diagnostic feature retained by the fossil valley dunes.

The other category of regional dune-rock deposits is that of the falling dune, always more remote from source materials and always in a leeward position sheltered by a geographic barrier. The most remote example of such a deposit is at Mesa Las Barracas, where a fossil dune sits at the end of a 4-km-long corridor to the nearest beach with a northern exposure. The highest barrier behind which a fossil dune is known to have accrued is on Isla Carmen, where a ridge upward of 200 m in elevation blocks a north-facing valley, yet has dune rock on its protected flank.

Dune rock in the lower Gulf of California is enriched in calcium carbonate derived from shell material, often more than 50 percent. High carbonate levels make dune rocks susceptible to erosion by dissolution that results in the formation of pinnacle karst. In the field, karst morphology is one of the best diagnostic features to look for when searching for dune rocks.

12 Active Geothermal Springs and Pliocene-Pleistocene Examples

Matthew J. Forrest and Jorge Ledesma-Vázquez

Geothermal springs occur in many locations throughout the Gulf Extensional Province, primarily along active faults (fig. 12.1). In some areas, such as Cerro Prieto and around the Tres Vírgenes, these geothermal springs are exploited as energy sources. In fact, the Cerro Prieto Geothermal Field is one of the oldest and most reliable geothermal fields in the world. It provides a significant proportion of the total power produced in the Baja California system, supplying a clean, renewable source of electricity for more than 3 million inhabitants. Geothermal springs also occur in coastal areas, particularly along the peninsular gulf coast, where fault systems associated with the opening of the Gulf of California typically involve geothermal activity (Libbey and Johnson 1997). Some geothermal springs extend into the intertidal and shallow-subtidal areas of the Gulf of California, where their unique geochemistry and high heat flow appear to affect the biological assemblages that surround them. This submarine activity is referred to as "hydrothermal venting." Examples of paleohydrothermal activity are also exposed throughout the Baja California peninsula, offering a unique opportunity to apply the principle of uniformitarianism, the observation that fundamentally the same geological processes that operate today also operated in the past.

Submarine hydrothermal venting occurs in diverse tectonically active settings throughout the world. Deep-sea hydrothermal venting is generally associated with active tectonic plate boundaries, such as slow- and fast-spreading centers, fracture zones, and back-arc spreading centers in subduction zones (German et al. 1995). Seafloor-spreading centers, where cold seawater circulates through the highly faulted crust and reacts with hot (> 300°C) basalts, are important components of the global chemical mass balance of the oceans over geological time, because high-temperature hydrothermal inputs play a significant role for the oceanic budgets of many dissolved metals

in seawater (Edmond et al. 1982). Deep-sea hydrothermal vents also support unique ecosystems that rely mainly on sulfide-rich hydrothermal fluids and chemosynthetic bacteria, rather than photosynthesis, to provide energy and nutrition to the animals that surround them. Within the Gulf of California, deep-sea hydrothermal vents occur at seafloor-spreading centers in the Guaymas basin and along the Alarcón Ridge (Lonsdale and Becker 1985; Fisher et al. 2001). These seafloor-spreading centers are related to the rifting of the Baja California peninsula away from North America and the opening of the Gulf of California (see fig. 1.1).

Shallow-water hydrothermal vents also occur throughout the world's marine environments (Tarasov et al. 2005). The shallow vents in the Gulf of California and western Mexico appear to be related to the extensional tectonic regime influencing this region (Prol-Ledesma 2003). Only three areas of shallow submarine hydrothermal venting in Mexico have been the target of extensive scientific studies: Punta Banda in Baja California (Vidal et al. 1978), Bahía Concepción in Baja California Sur (Prol-Ledesma et al. 2004; Forrest et al. 2005; fig. 12.1, BC), and Punta Mita near Puerto Vallarta, Mexico (Prol-Ledesma et al. 2002). Subtidal hydrothermal venting also has been reported from the Wagner basin, a spreading center in the northern Gulf of California that averages 80 m in water depth. Few published data concerning hydrothermal activity are currently available from this location, aside from scattered reports about helium anomalies and high heat flows (Suárez-Bosche et al. 2000). In addition to these subtidal sites, several intertidal hot springs have been investigated along the peninsular side of the Gulf of California near San Felipe, Puertecitos, and Bahía Concepción, mainly to assess whether they represent exploitable geothermal resources (Barragán et al. 2001). Intertidal hot springs also occur at Agua Verde and La Ventana and near

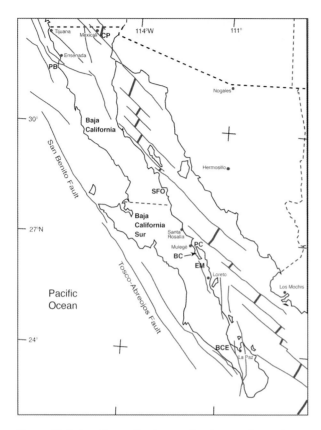

Figure 12.1. Locations of paleo- and active geothermal springs and hydrothermal vents within the Gulf Extensional Province including the Gulf of California. Refer to text for details. CP, Cerro Prieto; PB, Punta Banda; SFO, Bahía San Francisquito; BC, Bahía Concepción; PC, Peninsula Concepción; EM, El Mangle Block; BCE, Bahía Coyote.

Los Barriles on the East Cape, where the Buena Vista Resort (a famous fishing destination for decades) obtains all its water from a geothermal spring that runs under the hotel and continues offshore.

Despite the accessibility of shallow-water hydrothermal vents, the effects of hydrothermal venting on organisms found nearby have been poorly studied compared to deep-sea vents. The biogeochemical processes in zones of shallow-water hydrothermal activity may be more complicated than those occurring around deep-sea hydrothermal systems, because they are situated within the euphotic zone. Therefore, primary production from both chemosynthetic and photosynthetic sources may occur simultaneously, and bacteria, cyanobacteria, and microalgae often form dense mats around the shallow-water vents (Tarasov et al. 2005). Biodiversity at the microbial level may be an important driving force behind the high macrofaunal biodiversity also observed around shallow-water hydrothermal vents throughout the world. More food

and energy may be available for organisms to exploit in these settings than in the surrounding areas. Enhanced biodiversity and large numbers of fishes have been reported around shallow vents in the Gulf of California, suggesting that these areas represent biological hot spots.

Geochemistry of Fluids from Shallow Vents

The geochemistry of hydrothermal vent fluids can be very informative regarding the sources and pathways involved in the generation and migration of fluids. Shallow-water hydrothermal vents usually emit gas bubbles along with hot liquids due to a significant phase of exsolved gas. This activity has been referred to as "gasohydrothermal venting" (Tarasov et al. 1990). Gas bubbling can be quite vigorous in some areas (fig. 12.2), and the bubbles that rise to the surface offer valuable clues about the locations and extent of shallow-water hydrothermal vents. Researchers and fishermen, alike, follow such clues to locate areas of venting activity. Snorkeling or SCUBA diving among the bubbles and hot shimmering waters offers a unique and unforgettable experience to recreational divers, as well. Around shallow vents just offshore from the Caribbean island of Dominica this gasohydrothermal activity is billed as "champagne diving" by local dive-tour operators, and the shallow vents at Champagne Reef are the most famous diving and snorkeling sites on the island.

Most of the shallow-water hydrothermal vents that have been investigated within the Gulf of California and bordering the Baja California peninsula share similar water chemistries. Shallow vents at Punta Mita, Punta Banda, and Bahía Concepción (fig. 12.1)

Figure 12.2. Gasohydrothermal venting at 5-m water depth, Bahía Concepción. Photo credit: Roy E. Price

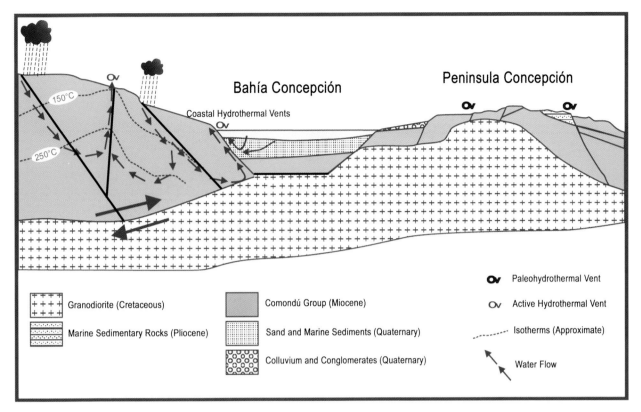

Figure 12.3. Geohydrological model of meteoric water circulation through faults at shallow-water hydrothermal vents in Bahía Concepción (modified after Prol-Ledesma et al. 2004).

discharge geothermal waters that are more dilute than seawater with lower concentrations of sodium, chlorine, and magnesium but higher concentrations of other elements, most notably silicon, calcium, manganese, and arsenic (Prol-Ledesma 2003). Detailed analyses of the geochemistry of these geothermal waters indicate that the shallow vents discharge meteoric (fresh) waters mixed with entrained seawater (Vidal et al. 1978; Barragán et al. 2001; Prol-Ledesma et al. 2004).

The permeability of the rocks on the Baja California peninsula is very low, confined mainly to areas where active faults occur (Portugal et al. 2000). Therefore, the formation and locations of coastal geothermal springs are controlled by infiltration and heating of meteoric waters, which are subsequently mixed with varying amounts of seawater along deep-seated faults related to the extensional tectonic structures associated with the opening of the Gulf of California (Barragán et al. 2001). This model is illustrated in figure 12.3 for the submarine hydrothermal vents in Bahía Concepción.

Elevated levels of arsenic and mercury have been measured in hydrothermal fluids and mineral precipitates from Punta Mita and Bahía Concepción (Prol-Ledesma et al. 2002; Canet et al. 2005a). These toxic elements seem to have little effect on the organisms surrounding the vents, which suggests that the arsenic and mercury are not readily bioavailable. The main mechanism for the reduction in bioavailability of arsenic around shallow vents appears to be due to the rapid incorporation of the arsenic into iron minerals that precipitate in areas of venting. These bright orange to yellow iron precipitates are invariably present on rocks and soft sediments surrounding hydrothermal vents (fig. 12.4) and may be responsible for locking up more than 95 percent of the arsenic from the fluids (Price and Pichler 2005). For example, around shallow-water vents offshore from Ambitle Island, Papua New Guinea, about 1,500 g of arsenic are being discharged per day via hydrothermal fluids flowing directly through a coral reef ecosystem (Pichler et al. 1999). Despite the high concentrations of arsenic, corals, clams, and fish show no obvious signs of response, and there is no evidence for accumulation of arsenic in their skeletal materials (Pichler et al. 1999; Price and Pichler 2005). Around the shallow vents in Bahía Concepción, arsenic-rich iron precipitates also

Figure 12.4. Mineral precipitates forming around gasohydrothermal vents at 5-m water depth in Bahía Concepción. Photo credit: Matthew J. Forrest

accompany mercury mineralization. The textures and fabrics of these mineral assemblages suggest they may be microbially mediated (Canet et al. 2005a).

Geological Framework for Shallow Vents at Bahía Concepción

Bahía Concepción represents one of the best examples of the extensional basins and accommodation zones that formed along the Baja California peninsula during the opening of the Gulf of California. Bahía Concepción's narrow, elongate shape results from a half-graben controlled by northwest–southeast trending faults (Johnson and Ledesma-Vázquez 2001). The most prominent of these faults make up the Concepción Fault Zone on Peninsula Concepción (McFall 1968), which acts as the eastern margin of the bay (fig. 12.5). The steep escarpment along much of the western shore of Bahía Concepción suggests that the western

margin of the bay also is bound by a fault zone with a northwest–southeast trend. Additional evidence of faulting is provided by active venting of hydrothermal fluids that occurs at several locations along the western shore (McFall 1968).

Most of the modern hydrothermal activity within Bahía Concepción occurs as discrete intertidal hot springs. Near Punta Santa Bárbara, however, copious discharge of gas bubbles and hydrothermal waters with temperatures as high as 92°C can be observed in multiple locations. Gasohydrothermal venting occurs through rocks and sediments ranging from the intertidal to 13-m water depths along a roughly linear trend extending over 750 m of coastline (fig. 12.5). Forrest et al. (2005) referred to the northwest–southeast trending faults that delineate the western shore of Bahía Concepción and control the modern hydrothermal activity as El Requesón Fault zone. This name derives from Isla El

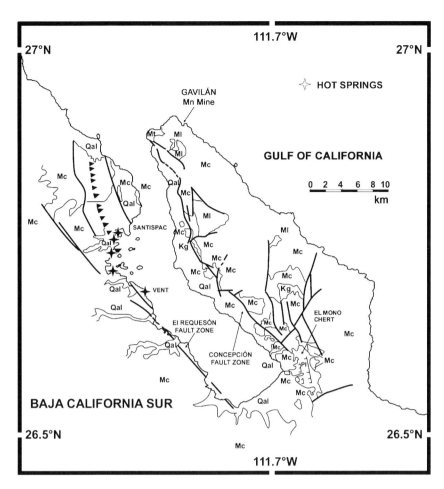

Figure 12.5. Geological map of the Bahía Concepción area. Thick lines represent faults. Kg, Cretaceous granitoids; Mc, Miocene Comondú Group; MI, Miocene undifferentiated intrusive; Mt, Miocene Tirabuzón Formation; Pl, Pliocene Infierno Formation; Qal, Quaternary alluvium. The line of black arrowheads marks the position of a low-angle fault with the falling block moving in the direction of the arrowheads. Modified from McFall (1968) and Ledesma-Vázquez and Johnson (2001).

Requesón, a prominent and conspicuous geographic feature attached to the western shore of Bahía Concepción by a 350-m long tombolo of rhodolith-derived carbonate sands. The word *requesón* refers to cottage cheese curds, undoubtedly an allusion to the lumpy, pustular appearance of crushed and bleached rhodolith debris, the main component of the carbonate sands on the tombolo and in the shallow bays adjacent to Isla Requesón (Hayes et al. 1993; chap. 6).

The geochemistry of the gases released at shallow-water hydrothermal vents provides important information about biogeochemical pathways and interactions, as well as the relationship of the venting activity with regional tectonic structures. Gas collected at the hydrothermal vent in Bahía Concepción was found by Forrest et al. (2005) to be composed mainly of nitrogen gas (53 percent), carbon dioxide (43 percent), and methane (2.2 percent). Carbon dioxide is the major component of gases from most of the shallow-water hydrothermal systems that have been sampled, but nitrogen and methane may also be present in significant concentrations, particularly at shallow vents off

northern Baja California and at Punta Mita on mainland Mexico (Vidal et al. 1978; Prol-Ledesma et al. 2002). High concentrations of nitrogen in gases are usually derived from thermal alteration of sedimentary organic matter (Jenden et al. 1988). The principal sources of methane in natural gases are either biogenic, due to microbial processes in low-temperature anaerobic environments, or thermogenic, where high temperatures (> 100°C) are sufficient to break chemical bonds in organic matter (Welhan 1988).

The carbon and nitrogen stable isotope values from the gas from Bahía Concepción suggest that it is primarily derived from the breakdown of organic matter within sediments (Forrest et al. 2005). Gases from other coastal hydrothermal systems in Mexico also show enrichment in nitrogen, with concentrations as high as 88 percent at Punta Mita (Prol-Ledesma et al. 2002) and 44 percent at Punta Banda (Vidal et al. 1978). Punta Mita and Punta Banda gas samples exhibited far lower concentrations of carbon dioxide (0.19 and 2.0 percent, respectively) than the gas from Bahía Concepción. The higher concentrations of carbon dioxide in the Bahía Concepción samples may reflect higher temperatures at depth in this system, as carbon dioxide can be expected to become the predominant gas in geothermal environments at temperatures higher than 180°C (Giggenbach 1997). Chemical geothermometers indicate a deep reservoir temperature of approximately 200°C for the vent fluids in Bahía Concepción (Prol-Ledesma et al. 2004; Forrest et al. 2005). Additionally, some of the carbon dioxide in the gas may be derived from hot, acidic geothermal fluids dissolving rhodolith matter within sediments. The δ^{13}C values from the rhodolith calcium carbonates within Bahía Concepción are similar to the δ^{13}C values of the carbon dioxide from the Bahía Concepción gas samples (D. Steller, personal communication 2004). Rhodoliths are not known to be present at Punta Mita or Punta Banda, but rhodoliths are very common in Bahía Concepción (see chap. 6). In fact, Diana Steller, who was working on rhodolith beds in the area, originally brought the subtidal hydrothermal vents in Bahía Concepción to our attention. The methane stable isotope values and methane-to-ethane ratios clearly indicate a thermogenic source for the gases collected from Bahía Concepción (Welhan 1988). The δ^{13}C values of the methane are consistent with derivation of the gas by thermal cracking of marine algal kerogens, or pseudokerogens (Forrest et al. 2005). Large quantities of rhodolith-derived sediments and the high productivity of phytoplankton in Bahía Concepción may provide the necessary algal pseudokerogens, which are subsequently exposed to geothermal fluids in excess of 100°C along conduits provided by El Requesón Fault zone. This results in the geologically instantaneous generation of nitrogen-, carbon dioxide–, and methane-rich gases.

The gases from the shallow-water hydrothermal vents at Punta Mita and Punta Banda have significantly higher concentrations of methane (11.8 and 51.4 percent, respectively, vs. 2.2 percent at Bahía Concepción), and the carbon stable isotopes of the methane are considerably lighter than the Bahía Concepción samples (Vidal et al. 1978; Prol-Ledesma et al. 2002). These data may indicate that the methane from Punta Mita and Punta Banda is derived from a mixed source, with contributions from biogenic (microbial) and thermogenic sources. In addition, at Punta Mita, there is evidence that carbonate mounds found around venting sites may have formed due to calcite precipitation induced by microbial methane oxidation (Canet et al. 2003).

Helium stable isotope ratios are particularly informative about the geological pathways involved in the formation of fluids at sites of hydrothermal activity because they offer a valuable tracer for the extent of mantle contribution. Mantle-derived gases show a high ^{3}He/^{4}He ratio relative to the atmosphere due to incomplete degassing of mantle ^{3}He following the condensation of Earth 4.5 billion years ago (Clarke et al. 1969). The contrast in ^{3}He/^{4}He ratios in crustal and magmatic systems provides a means of identifying a mantle input in a crustal system (Poreda et al. 1986). The measured ^{3}He/^{4}He ratio for the Bahía Concepción gas is 1.32 R_A (R_A is the ^{3}He/^{4}He ratio typically found in air: 1.4×10^{-6}), indicating a 16.3 percent mantle helium contribution to the Bahía Concepción gas sample (Forrest et al. 2005). Places where ^{3}He/^{4}He ratios > 1 R_A occur are associated with regions of active extension as a result of crustal thinning. In particular, boundary faults may serve as conduits for the mantle helium (Oxburgh and O'Nions 1987). Therefore, the mantle helium being released along El Requesón Fault zone supports assertions by Ledesma-Vázquez and Johnson (2001) that the El Requesón Fault zone may represent the master fault that led to the formation of the half-graben in the Bahía Concepción area.

Ecological Significance of Shallow-Water Vents

Deep-sea hydrothermal vents are characterized by macrofaunal biomasses so much greater than the surrounding benthos that they are often referred to as "oases" in the otherwise barren "deserts" of the deep seafloor (Carney 1994). This pattern is attributable to local primary productivity due to high chemosynthetic microbial activity. Chemosynthesis refers to the biosynthesis of organic compounds from carbon dioxide or methane using energy and reducing power from inorganic compounds such as hydrogen sulfide. Generally, sulfur-oxidizing chemosynthetic bacteria appear to be the major primary producers in hydrothermal environments, but methanotrophic bacteria may also be important (Conway et al. 1994). Some invertebrates around deep-sea vents harbor chemoautotrophic bacteria within their tissues, and these symbionts provide some or all of the nutritional needs of the host organisms. Invertebrates with symbionts have been called "vent obligate" organisms (Barry et al. 1996), because they cannot survive unless their symbionts have access to the necessary compounds within the vent fluids. Chemosynthetic bacteria are also present as microbial mats growing on hard substrates or sediments around shallow-water and deep-sea hydrothermal vents. Numerous organisms use these microbial mats as food sources, which may attract high numbers of predators to shallow vents. Although vent obligate macrofauna are abundant around deep-sea hydrothermal vents in the Gulf of California (Lonsdale and Becker 1985), they have not been found around the shallow-water hydrothermal vents in the region. However, observations suggest that both biodiversity and macrofaunal biomass are particularly high around these shallow vents, and this pattern has been observed at shallow vents throughout the world (e.g., Cocito et al. 2000; Tarasov et al. 1999).

The chemosynthetic nature of the base of the food webs at deep-sea hydrothermal vents was originally surmised from the unusual carbon and nitrogen stable isotopes of the vent fauna (Conway et al. 1994). Stable carbon and nitrogen isotope ratios have also been used to demonstrate that some animals foraging around a shallow-water hydrothermal vent near White Point, California, derive substantial nutrition from sulfur-oxidizing bacterial mats growing in areas of venting (Trager and De Niro 1990). Studies from the shallow vents in Bahía Concepción also revealed significant differences in the stable isotopes of animals foraging near the vents versus the same species collected away from the vents (Forrest 2004). The shallow vents in Bahía Concepción appear to lack the sulfur- and/or methane-oxidizing chemosynthetic bacterial mats commonly found around other hydrothermal vents. This is likely due to the fact that hydrogen sulfide is not present at detectable levels in the vent fluids, and because the small amount of methane present is generated by thermogenic sources. However, areas of active venting are invariably covered by a conspicuous layer of flocculent material, which is comprised of diatoms and other microalgae along with diverse assemblages of bacteria in a matrix of metals (fig. 12.4). It is likely that the precipitation of these metals (e.g., iron, manganese) and the mineralization of manganese-barium-mercury crusts found only in areas of active venting are mediated by microbes (Canet et al. 2005a).

True bacterial mats—likely comprised of sulfur-oxidizing and/or methanogenic and methanotrophic microbes—are present around the shallow vents at Punta Mita (Canet et al. 2003) and Punta Banda. The bacterial mats at Punta Banda have very negative stable isotope ratios, typical of chemosynthetic bacteria,

with a mean $\delta^{13}C$ value of –40.63‰ and a mean $\delta^{15}N$ value of –7.26‰. Because stable isotope values from the bacterial mats at Punta Banda are quite distinct from the typical marine photosynthetic values ($\delta^{13}C$ from –15 to –25‰; $\delta^{15}N$ from 2 to 12‰), the approximate amount of dietary contribution derived from the bacterial mats can be calculated using simple mixing models.

Several organisms appear to be incorporating significant amounts of vent bacterial mats into their diets, including an unidentified sponge and a small fish, the blackeye goby (*Rhinogobiops nicholsii*; M. Forrest, personal observations). These findings are quite significant, because they offer some of the first evidence that fish may obtain dietary contributions from chemosynthetic bacteria around shallow-water hydrothermal vents, which may help to explain why fishermen and scientists often report high abundances of fishes around shallow vents. Fishes around the Bahía Concepción vents have been observed picking at the flocculent materials forming in areas of active venting and foraging on the sponges and other filter-feeding invertebrates that are particularly abundant there. In addition, fishes are often observed hovering over the vents and "bathing" in the hydrothermal fluids. This behavior may reflect an attempt to remove parasites susceptible to the high concentrations of metals in the fluids and/or the elevated temperatures. It is also possible that fishes may find optimal temperatures for growth and successful reproduction by exploiting the temperature gradients available around the vents.

Infaunal organisms are scarce in areas of venting through soft sediments when compared with areas not affected by the venting in Bahía Concepción (Forrest 2004). It is likely that the high temperatures (up to 90°C) and acidity (pH as low as 6.2 vs. about 8.0 for seawater) of the vent fluids prevent the worms, bivalves, and small crustaceans that typically live within the sediments from surviving in areas of venting. However, these stressful environments may select for more mobile organisms. This appears to be the case for gastropods in the genus *Nassarius*, which are commonly observed around vent sites in Bahía Concepción. Nassariid gastropods typically are scavengers that feed on dead and decaying animals, and they are capable of fairly rapid movement. These gastropods are also abundant around shallow vents in Milos, Greece (Southward et al. 1997), and in Matupi Harbour in Papua New Guinea (Tarasov et al. 1999). The fact that nassariid gastropods have been found in abundance around shallow-water vents occurring in disparate settings throughout the world suggests that they should be good indicator species in terms of interpreting potential areas of paleohydrothermal activity.

Paleohydrothermal Vents

There is a very thin line that divides active hydrothermal vents from paleohydrothermal vents surrounding the Gulf of California, because most of the tectonic structures that are associated with hydrothermal activity have been active at one time or another. The peninsular fault systems associated with the gulf (chap. 1) typically involve hydrothermal activity, and there is ample evidence of paleohydrothermal activity throughout the gulf region. Many localities exhibit deposits associated with faults that acted as the conduits for the hydrothermal fluids. In most cases, these fluids interacted with the host rock, altering the original lithology or generating associated deposits. This characteristic allows for initial detection by analysis of satellite images for large regions, followed up by fieldwork.

Paleohydrothermal Activity near Bahía Concepción

There are several lines of evidence that extensive paleohydrothermal activity occurred along the Concepción Fault Zone. McFall (1968) concluded that manganese veins exploited at the Gavilán mine on Peninsula Concepción (fig. 12.5) were of hydrothermal origin. Several faults cut across Pliocene strata on a trend parallel to the Concepción Fault zone at the base of the Peninsula Concepción (fig. 12.5). These faults acted as conduits for hydrothermal fluids, resulting in paleohydrothermal vent sites (Johnson et al. 1997; Ledesma-Vázquez et al. 1997). One paleohydrothermal vent site that occurs along a major fault line is located on the northeastern side of Cerro Prieto, where red mudstone shows evidence of peculiar disruption reminiscent of gas bubbles rising through fine sediments. Preservation of this feature in Upper Pliocene strata means that fault-fed hydrothermal activity occurred during the Late Pliocene (Johnson et al. 1997). This site appears to offer an excellent example of the gasohydrothermal activity occurring at modern shallow-water hydrothermal areas described earlier in this chapter.

Another remarkable site of fault-controlled paleohydrothermal activity on the Peninsula Concepción

is exemplified by a member-level stratigraphic unit named El Mono (fig. 12.5), which features nearly 14 m of bedded cherts including dense, competent layers of brown or white chert interbedded with soft, poorly consolidated layers of finely brecciated chert (Ledesma-Vázquez et al. 1997). Individual chert beds within El Mono Member range in thickness from a few centimeters to a maximum of 4.5 m. Bioturbation is noticeable in some chert layers, and the presence of *Ophiomorpha* ichnofossils (branching burrows made by shrimp-like crustaceans) indicates that deposition took place in the shallow intertidal to upper intertidal zones. Occurrences of in situ root casts belonging to a species of the black mangrove *Avicenia* corroborate the coastal setting of the deposit. A Late Pliocene age is suggested by the occurrence of the index fossil *Clypeaster marquerensis*, a sand dollar commonly found in the overlaying limestone unit (Ledesma-Vázquez et al. 1997).

The repetitive sequence of reworked and massive chert beds from El Mono Member formed in a shallow-marine environment most likely influenced by a high rate of evaporation. Such evaporation would have increased water salinity, which in turn promoted silica precipitation. Salinity increased to the point that halite precipitated along with silica in some of the beds, but the basin did not completely evaporate, as indicated by the absence of saline crusts or evaporites deposits. Water in the basins was shallow enough to allow wave action and erosive processes that reworked some of the chert beds and introduced small amounts of terrigenous materials.

Evidence of post-Pliocene hydrothermal activity along basin-cutting faults of the Concepción Peninsula suggests that hydrothermal fluids rose through the underlying silica-rich volcanic rocks of the Comondú Formation. Silica dissolved from the Comondú Formation was deposited in the overlying Infierno Formation. This is similar to the process suggested by Siever (1983) for rift valleys with initial high heat flow. Once the already silica-rich hydrothermal solutions entered the basin, they attacked and dissolved some of the calcareous layers, thereby raising the pH of the originally acidic solutions. As pH lowered with the dissolution of silica, the saturated hydrothermal solutions began to redeposit secondary silica in the openings left by earlier solution activity. The newly deposited silica took the form of opal-A, which was transformed to opal-CT as the region continued to experience heating from hydrothermal sources. Work by Iijima (1988) on the diagenetic transformation of volcanic ash to opal-A, and then on to opal-CT, presents a similar sequence of minerals. The same reaction pathways invoked by Iijima (1988) are suggested to be responsible for the transformation that took place at El Mono. An important difference, however, is that according to Iijima, depth of burial provided sufficient heat to drive the reaction, whereas hydrothermal fluids percolating through faults provided the necessary heat at the site of the Mono cherts. Hot springs within a mangrove stand at Santispac, Bahía Concepción (fig. 12.5), serve as a modern-day model for the paleohydrothermal activity that was involved in the formation of the Mono cherts.

Other interesting deposits associated with shallow-water hydrothermal vents include fossil and modern silica-carbonate stromatolites. Finely laminated silica-carbonate hot spring deposits are present around the modern shallow-water hydrothermal vents in Bahía Concepción, as illustrated by Canet et al. (2005b), who also described a 75-m-long fossil bed of silica-carbonate stromatolites cropping out along a cliff next to the active vent area. The morphology of the fossil hot spring deposits is variable; they form small columnar, bulbous, and smooth undulating microstromatolites up to 10 cm thick. The stromatolites occur irregularly upon the volcanic bedrock, frequently incorporating clastic fragments. The modern hot-spring deposits form crusts and coalescing rims over volcanic pebbles and boulders. Around the main hot-spring outflow conduits, the stromatolites coat a structureless aggregate formed by allochthonous material cemented by silica with minor amounts of calcite, barite, and manganese oxides (Canet et al. 2005b).

Stromatolites represent the oldest known fossils on Earth, dating back nearly 3.5 billion years. They were constructed in shallow-marine environments by the activities of microbes trapping and binding sediments and mediating mineral precipitation. Some interpretations suggest they may have formed in hydrothermal settings. Much of what we understand about the existence of early life forms comes from the examination of siliceous microfossils in deposits such as the 3.5-billion-year-old Apex cherts in Australia (Konhauser et al. 2003). Structures resembling cyanobacteria were observed in the Apex cherts (Schopf et al. 2002). These structures were reinterpreted as secondary artifacts created in a hydrothermal setting; however, carbon-isotope values do imply a significant biological contribution, perhaps from hyperthermophilic bacteria (Brasier et al. 2002). The silica-carbonate stromatolites in Bahía Concepción are rare examples

of modern stromatolites from a marine hydrothermal setting. Microbial remains are present in both modern and fossil stromatolites, and the fabrics and textures of the deposits indicate that precipitation is mediated and constrained by the biological activities of microbes (Canet et al. 2005b). In light of the controversy surrounding the extent and type of biological activity involved in creating Archaean stromatolite-like clasts, a detailed examination of the Bahía Concepción stromatolites may provide more insight into the formation of siliceous microfossils in similar settings.

The Boleo Mining District at Santa Rosalía

The mining town of Santa Rosalía (fig. 12.1), located approximately 100 km north of Bahía Concepción, is famous for its stratiform nonmarine and marine deposits with high concentrations of copper, manganese, and cobalt that mineralized within closely related basins. The basins initially formed in response to early extensional rifting in the gulf region sometime prior to about 10 Ma. The mineralization of these deposits resulted from hydrothermal fluids that ascended along basin faults (Conly et al. 2006). Sediments were laid down as units called the Tirabuzón (Carreño 1981), Infierno, and Santa Rosalía Formations within a series of sub-basins (Wilson and Rocha 1955). Attributed collectively to the Santa Rosalía basin, this succession of nonmarine and marine formations is now exposed as an uplifted wedge with a maximum thickness of 500 m covering an area extending tens of kilometers north and south of Santa Rosalía and 10 km inland from the coast. Ridges formed by the tilted edges of basement rocks divide the various sub-basins. At the Boleo sub-basin, two subparallel but discontinuous basement highs are the locus of mineralization. One occurs near the gulf coast and the other is located from 3 to 5 km inland.

The Boleo sub-basin was partially filled with terrestrial sediments eroded from highlands to the west and by distally erupted volcanic ash. At the Saturno trench in the mining district, it is possible to see an elongated feature formed by silicified, clay-rich breccia with inclusions of jasper, chrysocolla, cryptomelane, limonite, and other related minerals. This particular site has been identified as a paleohydrothermal vent, and mineralization is localized along a listric fault. Metal-rich accumulations in the Boleo sub-basin appear to have resulted under the influence of a hydrothermal system that operated under similar conditions to those found today in the adjacent deep-water Guaymas basin, although the Boleo deposit also entailed enrichment of sedimentary units including Pliocene alluvial deposits. Evidence of hydrothermal activity found along basin-cutting faults indicates that hot water enriched with mineral ions rose through the silica-rich volcanic rocks of the underlying Miocene Comondú Formation and subsequently mineralized the accumulating sedimentary units (Conly et al. 2006).

Other Sites Showing Paleohydrothermal Activity

A small north–south oriented tectonic block, named El Mangle Block by Johnson et al. (2003), is located on the eastern flank of the Cerro Mencenares volcanic complex, which is centered 30 km north of Loreto (fig. 12.1). The block is configured as a 7-×-1-km horst separated from Cerro Mencenares by a north–south valley fault that formed during the extensional phase of tectonics in the gulf sometime prior to about 3.5 Ma. The block's southern end is defined by a normal fault striking N55°W parallel to the Atl Fracture in the modern Gulf of California. Named El Coloradito Fault (Johnson et al. 2003), this feature truncates an undisturbed carbonate ramp sequence adjoined to the south, which includes a tuff bed yielding a K/Ar age of 3.3 ± 0.5 Ma. Thus, El Coloradito Fault was active sometime after that date, but fits with the overall change in regime to transtensional tectonics initiated in the Gulf of California at about 3.5 Ma. A succession of smaller faults that run parallel to El Coloradito Fault also slice El Mangle Block at close intervals and hydrothermal alteration is extensive along all these oblique faults. Thick beds with abundant fossil pectens drape parts of the block from elevations as high as 100 m down to sea level. It is noteworthy that the pecten beds are heavily tectonized and mineralized by coatings with a green discoloration, which clearly owe their origin to hydrothermal activity.

DeDiego-Forbis et al. (2004) described another locality displaying similar paleohydrothermal activity along a narrow cliff within the coastal plain of Bahía Coyote, north of La Paz (fig. 12.1). They found locations near small normal faults, where massive *Porites* coral beds, molluscan fossils, and the sediment matrix are stained green by material remobilized from the underlying Cerro Colorado. Typically, the corals are weathered, coated with a dark-colored patina, and poorly preserved. The silica-coated *Porites* appear to be more common in the vicinity of the green-stained

deposits. Based on their field observations, DeDiego-Forbis et al. (2004) interpreted the affected areas as ancient hydrothermal vents located along faults. They also indicated that such vents are known in the present gulf, as described for Bahía Concepción. The authors also found that fossilized remains of *Nassarius tiarula* were abundant in the surrounding areas, which is interesting considering the observations that nassariid gastropods occur in abundance around shallow-water vents within Bahía Concepción and elsewhere throughout the world.

At Bahía San Francisquito (fig. 12.1, SFO), Pliocene strata occupy 10 km^2 in an embayment on the peninsular coast at about 75 km south of Isla Angel de la Guarda (see fig. 3.8). Boundary faults inside the basin were active during flooding but also remained active sometime afterward. As evidence, Pliocene limestone dips away from the northern paleoshore at inclinations internally adjusted between 20° and 25°. High dip angles suggest that the ramp was steepened after deposition, but syndepositional steepening also occurred during ramp accretion on the basis of changing dip angles preserved internally. Additional evidence for reactivation comes from localized hydrothermal deposits near the boundary faults. Fossilized aggregations of the bivalve *Glycymeris maculata* are widely found nestled in growth position among cobbles and boulders on the western side of the basin, and this bivalve also is associated with several mounded paleohydrothermal vents in the vicinity.

A Call for Conservation

Modern and fossil examples of shallow-water hydrothermal vents occur throughout the Baja California peninsula. These sites offer valuable clues about the processes involved in the extensional tectonics that created the Gulf of California and continue to shape the region to this day. Paleohydrothermal vents are associated with valuable mineral deposits that have led to lucrative mining operations. Invaluable information regarding the history of the biogeochemical processes related to the high biodiversity within the Gulf of California also may be mined from these sites. Modern shallow vents are highly productive and provide valuable resources to local fishermen. We still are merely scratching the surface in terms of gaining an understanding about the complex and fascinating biogeochemical interactions that occur around shallow-water hydrothermal vents in the Gulf of

California, and it is clear these areas are important for a myriad of reasons. Fishermen catch many commercially important species of fish and shellfish around the shallow vents in Bahía Concepción. Unfortunately, illegal fishing of endangered and threatened species also occurs regularly around shallow-water vents. Sea turtles are often observed around the shallow vents in Bahía Concepción. Large nets deployed around the vents and discarded shells found on the shore provide ample evidence for illegal exploitation of these magnificent creatures.

In the northern Gulf of California, fishermen report that totoaba (*Totoaba macdonaldi*) are being caught illegally around shallow-water hydrothermal vents. *Totoaba macdonaldi* is an endangered species endemic to the Gulf of California. The fish formerly reproduced in huge aggregations in the estuaries of the Colorado River. It has been demonstrated that totoaba grow and become sexually mature twice as fast if they spend a large part of their juvenile stage in brackish waters (Rowell 2006). Catches of thousands of tons of totoaba were recorded between 1930 and 1940, but the fishery abruptly crashed due to a devastating combination of overfishing and loss of natal habitat due to the damming of the Colorado River (Cisneros-Mata et al. 1995). In 1975, the Mexican Government declared an indefinite moratorium on fishing totoaba. The following year they were listed under the Convention on International Trade in Endangered Species. Totoaba are known to congregate around shallow vents in the northern gulf, perhaps because they find the lower salinities needed for optimal growth and reproductive success. It is also possible that the totoaba enjoy access to greater food availability around the shallow vents, as was found for the gobies around the Punta Banda vents. It is unclear what the relationship is between this fish and hydrothermal activity, but the fact remains that significant numbers are present around these shallow vents and that the surrounding areas have long been considered an important refuge for totoaba (M. Roman, personal communication 2007). In fact, Conal True, the director of an aquaculture facility for *T. macdonaldi* at Universidad Autónoma de Baja California in Ensenada, collects the fish that he uses to breed in the vicinity of shallow-water vents. If a connection does exist between the totoaba and shallow-water hydrothermal vents, protecting these sites by restricting fishing in the area would be advisable.

Recently, there has been some discussion about exploiting the energy potential that active shallow-water hydrothermal vents in the Gulf of California could provide. We feel such endeavors may be premature because the ecological significance of shallow-water hydrothermal vents remains poorly studied and understood. Recent research has revealed that unique heat-resistant microbes often are present around these shallow vents (Stetter 2006). Taq polymerase, the thermostable enzyme that is used in polymerase chain reaction to isolate and amplify a fragment or sequence of interest of DNA, was first isolated from *Thermus aquaticus*, a bacterium that lives in hot springs and hydrothermal vents. Polymerase chain reaction has become such an important tool to molecular biology that patents for the process were sold for $300,000,000. In 1989, Taq polymerase was the first ever "Molecule of the Year" named by *Science* magazine. In addition, the high arsenic concentrations within fluids from shallow vents allow researchers to derive important information about how microbes cycle arsenic through the ecosystem. Arsenic contamination of drinking water is a major problem throughout the world, and shallow-water vents represent natural laboratories where researchers can study ways in which microbial interactions help catalyze chemical reactions that may alleviate arsenic contamination (Pichler et al. 2006; Reed 2006). Recently, shallow-water volcanic carbon dioxide vents offshore from Ischia, Italy, were also used to investigate the potential future ecosystem consequences of ocean acidification due to anthropogenic additions of carbon dioxide to the atmosphere (Hall-Spencer et al. 2008). This study provided the first in situ insights into how shallow water marine communities might change when susceptible organisms are removed due to ocean acidification, further emphasizing the value of shallow-water vents as natural laboratories.

Because shallow-water hydrothermal vents offer far easier access than their deep-sea counterparts, they represent a potential goldmine of information and resources in the near future. Exposed sites of paleohydrothermal activity also are in danger of being lost or damaged as unchecked development continues to encroach upon formerly pristine areas throughout the Baja California peninsula. These sites are far too valuable to risk losing them forever, and they need to be protected so future generations can enjoy them and continue to explore their significance.

The following pages and enclosed CD are an extension of the *Atlas of Coastal Ecosystems in the Western Gulf of California*. Twenty-six high-resolution satellite images and two maps showing the area of the gulf coast of Baja California covered by the images are reproduced here. They can also be found on the CD in the folder labeled "Baja Atlas ASTER Images." These images (JPEG format) may be viewed using the standard image viewer provided by most computer systems. Due to the size of the images, downloading the image files to your computer desktop or hard drive will allow you to access the images more quickly.

The images on the CD are provided for your enjoyment and as an aid to research. They may not be reproduced for publication or used for any commercial purpose without the specific written consent of the University of Arizona Press.

Previewing the Images

As an aid to looking through the images on the CD, two galleries that require a Web browser are provided in the folder labeled "Preview."

To preview images that cover the state of Baja California, double-click the Web page (.htm) labeled "BajaCalifornia Gallery."

To preview images that cover the state of Baja California Sur, double-click the Web page (.htm) labeled "BajaCaliforniaSur Gallery."

About the Images

The images reproduced here and on the CD are derived from data collected by the Advanced Spaceborne Thermal Emission and Reflection Radiometer (ASTER) that is carried by the TERRA satellite. The scenes were constructed using bands 8, 3, and 1, which sample parts of the visible, near infrared, and shortwave infrared areas of the energy spectrum. The resulting images have been enhanced to improve the contrast and presentation of the original spectral data using the ENVI (version 4.3) program. The images were produced in the Remote Sensing and Geographic Information Systems Lab of Williams College. Information on the ASTER system is available at http://asterweb.jpl.nasa.gov/.

All data sets were collected by the NASA ASTER Program at the U.S. Geological Survey in Sioux Falls, South Dakota. Specific information about the data sets and the collection dates accompany each scene.

About the Web-based Galleries

The image preview galleries on the CD were created through the modification of output from the Adobe Photoshop (v. 7.0) Web Photo Gallery generator.

David H. Backus
Williams College, March 27, 2008

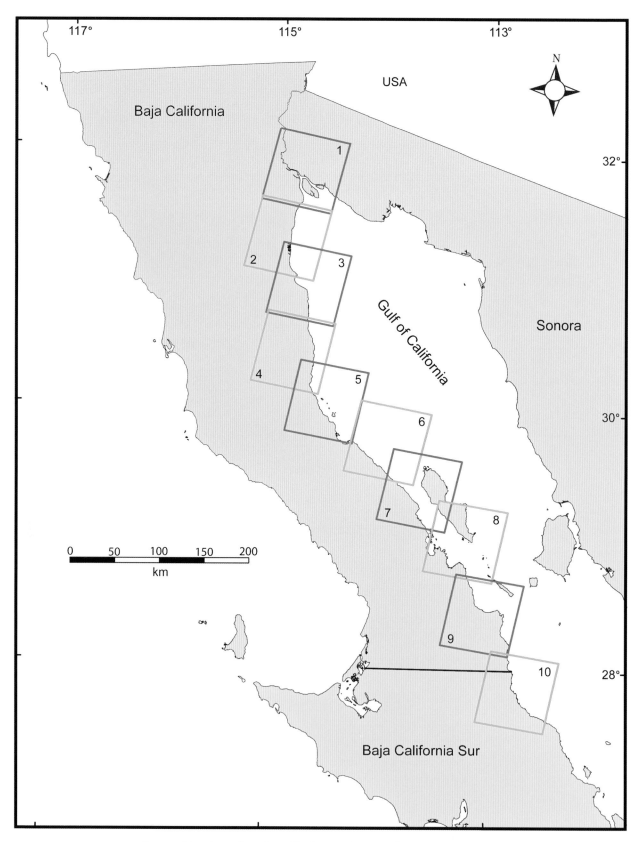

Geographic orientation of satellite images from northern Baja California.

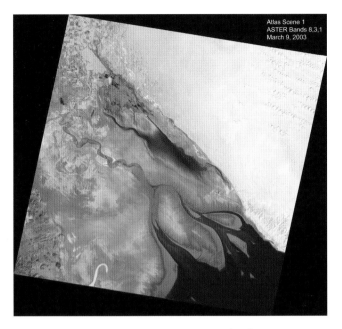

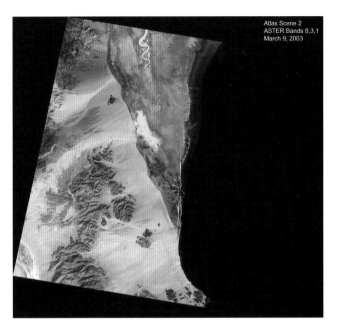

1. AST L1B.003.2017048587, 03/09/2003

2. AST L1B.003.2012092416, 03/09/2003

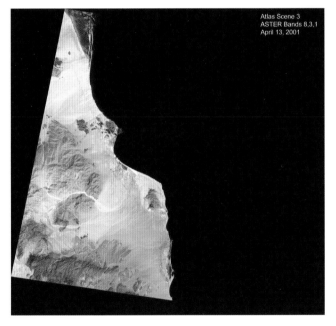

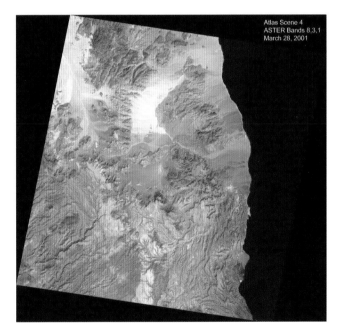

3. AST L1B.003.2019598476, 04/13/2001

4. AST L1B.003.2006051845, 03/28/2001

5. AST L1B.003.2018792928, 11/16/2001

6. AST L1B.003.2025986821, 09/28/2004

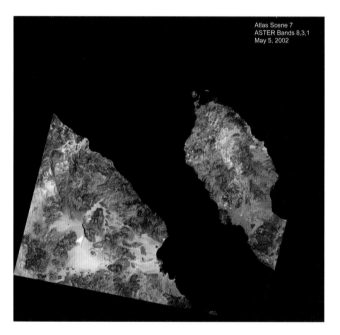

7. AST L1B.003.2007101582, 05/27/2002

8. AST L1B.003.2019036158, 01/12/2002

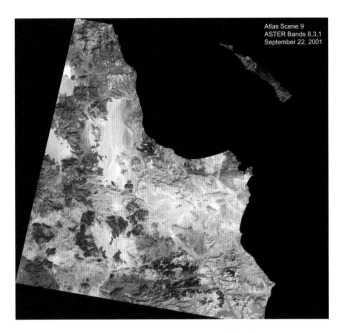

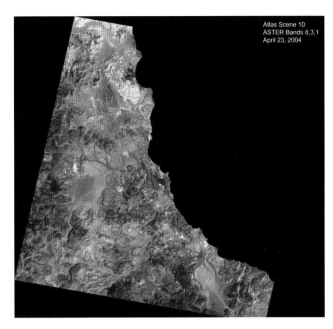

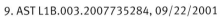

9. AST L1B.003.2007735284, 09/22/2001

10. AST L1B.003.2022776081, 04/23/2004

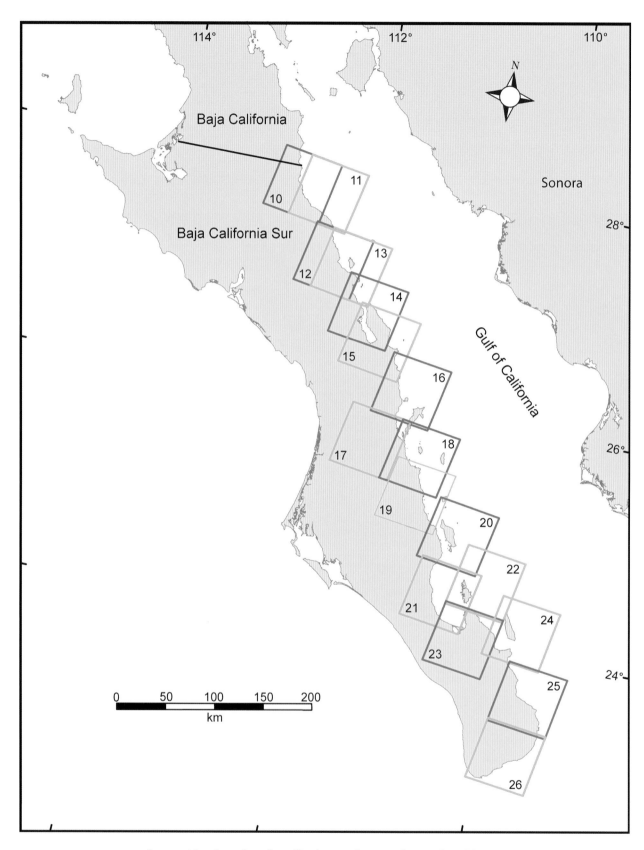

Geographic orientation of satellite images from southern Baja California.

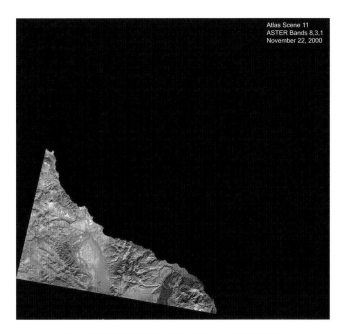

Atlas Scene 11
ASTER Bands 8,3,1
November 22, 2000

11. AST L1B.003.2020703320, 11/22/2000

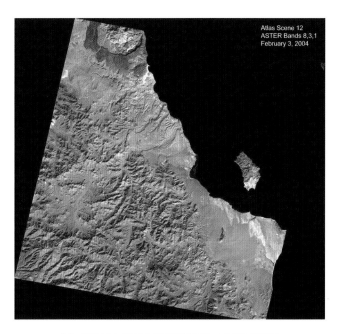

Atlas Scene 12
ASTER Bands 8,3,1
February 3, 2004

12. AST L1A.003.2021778853, 02/03/2004

Atlas Scene 13
ASTER Bands 8,3,1
May 25, 2004

13. AST L1B.003.2024093193, 05/25/2004

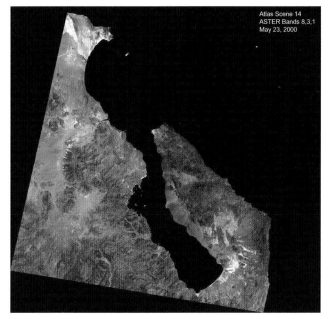

Atlas Scene 14
ASTER Bands 8,3,1
May 23, 2000

14. AST L1B.003.2018969828, 06/05/2002

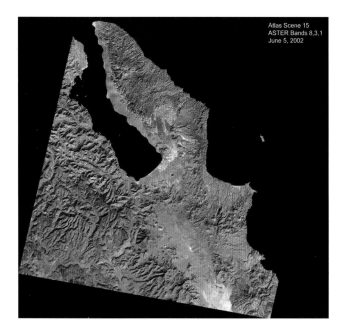

Atlas Scene 15
ASTER Bands 8,3,1
June 5, 2002

15. AST L1B.003.2017048587, 03/09/2003

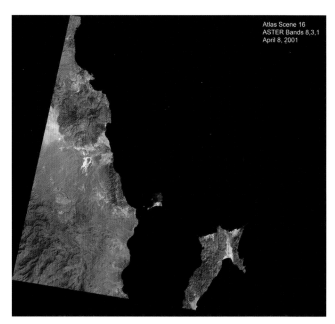

Atlas Scene 16
ASTER Bands 8,3,1
April 8, 2001

16. AST L1B.003.2017881231, 04/08/2001

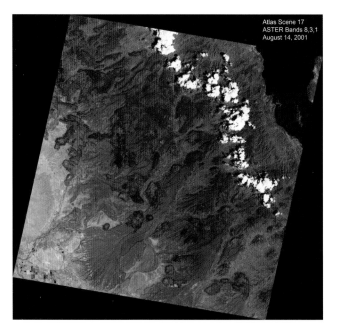

Atlas Scene 17
ASTER Bands 8,3,1
August 14, 2001

17. AST L1B.003.2003866371, 08/14/2001

Atlas Scene 18
ASTER Band 8,3,1
March 23, 2001

18. AST L1B.003.2014196273, 03/23/2001

Atlas Scene 19
ASTER Bands 8,3,1
May 14, 2005

19. AST L1B.003.2029047271, 05/19/2005

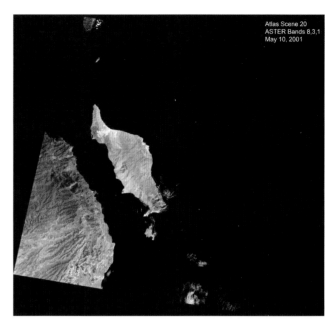

Atlas Scene 20
ASTER Bands 8,3,1
May 10, 2001

20. AST L1B.003.2003093452, 05/10/2001

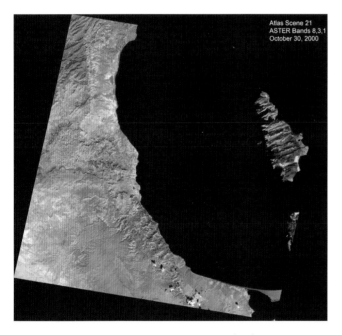

Atlas Scene 21
ASTER Bands 8,3,1
October 30, 2000

21. AST L1B.003.2018315496, 10/30/2000

Atlas Scene 22
ASTER Bands 8,3,1
June 7, 2000

22. AST L1B.003.2007178561, 06/07/2000

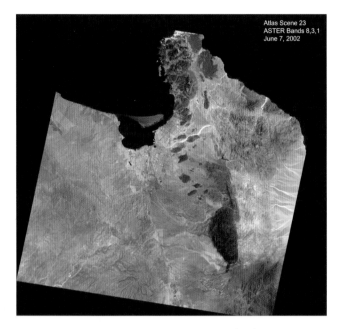

Atlas Scene 23
ASTER Bands 8,3,1
June 7, 2002

23. AST L1B.003.2007194562, 06/07/2002

Atlas Scene 24
ASTER Bands 8,3,1
May 7, 2005

24. AST L1A.003.2028941170, 05/07/2005

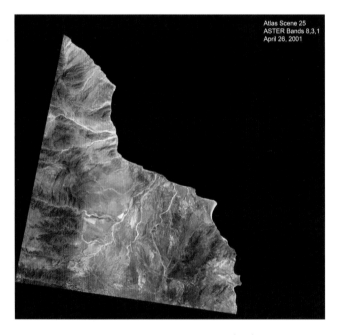

Atlas Scene 25
ASTER Bands 8,3,1
April 26, 2001

25. AST L1B.003.2020011242, 04/26/2001

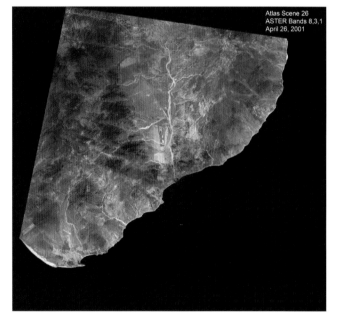

Atlas Scene 26
ASTER Bands 8,3,1
April 26, 2001

26. AST L1B.003.2020012060, 04/26/2001

Adey, P. J., and Lebednick, P. A. 1967. *Catalog of the Foslie Herbarium.* Tondheim, Norway, De Kongelig Norse Videnskabers Selskab Museet.

Adey, W. H., and Macintyre, I. G. 1973. Crustose coralline algae: a re-evaluation in the geological sciences. *Geological Society of America Bulletin* 84, 883–904.

Aguilar-Rosas, L. E., Aguilar-Rosas, R., Mendoza-González, A. C., and Mateo-Ciddoi, L. E. 2000. Marine algae from the northeast coast of Baja California, México. *Botanica Marina* 43, 127–139.

Aguirre, J., Riding, R., and Braga, J. C. 2000. Diversity of coralline red algae: origination and extinction patterns from the Early Cretaceous to the Pleistocene. *Paleobiology* 26, 651–667.

Alburto-Oropeza, O., and Balart, E. F. 2001. Community structure of reef fish in several habitats of a rocky reef in the Gulf of California. *Pubblicazioni della Stazione Zoologica di Napoli Marine Ecology* 22, 283–305.

Allison, E. C., Durham, J. W., and Mintz, L. W. 1967. New southeast Pacific echinoids. *Occasional Papers of the California Academy of Sciences* 62, 1–23.

Alvarez-Borrego, S. 1983. Gulf of California. In *Ecosystems of the World*, vol. 26, *Estuaries and Enclosed Seas*. Edited by B. Ketchum. New York, Elsevier Scientific, pp. 427–449.

Alvarez-Borrego, S., and Gaxiola-Castro, G. 1988. Photosynthetic parameters of northern Gulf of California phytoplankton. *Continental Shelf Research* 8, 37–47.

Alvarez-Borrego, S., and Lara-Lara, J. R. 1991. The physical environment and primary productivity of the Gulf of California. In *The Gulf and Peninsular Province of the Californias*. Edited by J. P. Dauphin and B. R. T Simoneit. American Association of Petroleum Geologists Memoir 47, pp. 555–567.

Alvarez-Borrego, S., and Schwartzlose, R. A. 1979. Water masses of the Gulf of California. *Ciencias Marinas* 6, 43–63.

Alvarez-Filip, L., and Reyes-Bonilla, H. 2006. Comparison of community structure and functional diversity of fishes at Cabo Pulmo coral reef, western Mexico, between 1987 and 2003. *Proceedings of the 10th International Coral Reef Symposium* 1, 216–225.

Alvarez-Filip, L., Reyes-Bonilla, H., and Calderón-Aguilera, L. E. 2006. Community structure of reef fishes in Cabo Pulmo reef, Gulf of California. *Marine Ecology* 27, 253–262.

Anderson, C. A. 1950. Geology of islands and neighboring land areas. In *1940 E. W. Scripps Cruise to the Gulf of California*. Part 1. By C. A. Anderson et al. Geological Society of America Memoir 43. 53 p.

Anonymous. 2001. Norma Oficial Mexicana NOM-059.

Anonymous. 2007. Decreto por el que se declara área natural protegida, con la categoría de Parque Nacional exclusivamente la zona marina del Archipiélago de Espíritu Santo, ubicado en el Golfo de California, frente a las costas del Municipio de La Paz, Baja California. *Diario Oficial de la Federación*, Mayo 10 2007. pp. 1–7.

Argote, M. L., Amador, A., Lavín, M. F., and Hunter, J. R. 1995. Tidal dissipation and stratification in the Gulf of California. *Journal of Geophysical Research* 100(C8), 16103–16118.

Arizpe-Covarrubias, O. 2005. El turismo como alternativa a la pesca en el manejo de un arrecife coralino. Caso Cabo Pulmo, Golfo de California. In *El manejo costero en México*. Edited by E. Rivera-Arriaga, G. J. Villalobos-Zapata, I. Azuz-Adeath, and F. Rosado-May. Universidad Autónoma de Campeche/SEMARNAT/CETYS/ Universidad de Quintana Roo, pp. 573–588.

Ash, J. E., and Wasson, R. J. 1983. Vegetation and sand mobility in the Australian desert dunefield. *Zeitschrift für Geomorphologie, Supplementbände*, 45, 7–25.

Ashby, J. R., Ku, T. L., and Minch, J. A. 1987. Uranium series ages of corals from the upper Pleistocene Mulege terrace, Baja California Sur, Mexico. *Geology* 15(2), 139–141.

Avila-Serrano, G. E., Flessa, K. W., Téllez-Duarte, M. A., and Cintra-Buenrostro, C. 2006. Distribution of the intertidal macrofauna of the Colorado River delta, northern Gulf of Colorado, Mexico. *Ciencias Marinas* 31, 649–661.

Axen, G. J. 1995. Extensional segmentation of the Main Gulf Escarpment, Mexico and United States. *Geology* 23, 515–518.

Axen, G. J. 2000. Rigid microplates, deforming terrains, and the slip budgets of the southern San Andreas Fault system and the Gulf of California. *Geological Society of America Abstracts with Programs* 32(7), 157.

Backus, D. H., Doctor, C., and Johnson, M. E. 2007. The remote sensing of sand dunes in Baja California, Mexico. *Geological Society of America Abstracts with Program* 39(10), 41–42.

Badan-Dangon, A., Koblinsky, D. J., and Baumgartner, T. 1985. Spring and summer in the Gulf of California: observation of surface thermal patterns. *Oceanologica Acta* 8, 13–22.

Bales, R. J., Christofferson, J. E., Escandon-Valle, F. J., and Peatfield, G. R. 2001. Sediment hosted deposits of the Boleo copper-cobalt-zinc district, Baja California Sur, Mexico. In *New Mines and Discoveries in Mexico and Central America*. Edited by T. Albinson and C. E. Nelson. Society of Economic Geologists Special Publication 8, pp. 291–306.

Barragán, R. R. M., P. Birkle, M. E. Portugal, G.V. M. Arellano, and R. J. Alvarez, 2001. Geochemical survey of medium temperature geothermal resources from the Baja California Peninsula and Sonora, México. *Journal of Volcanology and Geothermal Research* 110, 101–119.

Barry, J. P., Greene, H. G., Orange, D. L., Baxter, C. H., Robison, B. H., Kochevar, R. E., Nybakken, J. W., Reed, D. L., and McHugh, C. M. 1996. Biologic and geologic characteristics of cold seeps in Monterey Bay, California. *Deep-Sea Research I* 43(10–11), 1739–1762.

Bauer, B. O., and Davidson-Arnott, R. G. D. 2002. A general framework for modeling sediment supply to coastal dunes including wind angle, beach geometry, and fetch effects. *Geomorphology* 49, 89–108.

Baynes, T. W. 1999. Factors structuring a subtidal encrusting community in the southern Gulf of California. *Bulletin of Marine Science* 64, 419–450.

Beal, C. H. 1948. *Reconnaissance of the Geology and Oil Possibilities of Baja California, Mexico.* Geological Society of America Memoir 31. 138 p.

Bernardi, G., Findley, L. T., and Rocha-Olivares, A. 2003. Vicariance and dispersal across Baja California in disjunct marine fish populations. *Evolution* 57, 1599–1609.

Bigioggero, B., Capaldi, G., Chiesa, S., Montrasio, A., Vezzoli, L., and Zanchi, A. 1988. Post-subduction magmatism in the Gulf of California: the Isla Coronado (Baja California Sur, Mexico). *Instituto Lombardo (Rendiconti Scienze)* B121, 117–132.

Bigioggero, B., Chiesa, S., Zanchi, A., Montrasio, A., and Vezzoli, L. 1995. The Cerro Mencenares volcanic center, Baja California Sur: source and tectonic control on postsubduction magmatism within the gulf rift. *Geological Society of America Bulletin* 107, 1108–1122.

BIOMAERL. 1998. Maerl grounds: habitats of high biodiversity in European seas. In *Third European Marine Science and Technology Conference, Project Synopses*, Vol. 1: *Marine Systems*. Lisbon, European Commission, pp. 170–178.

Birkeland, C. 1977. The importance of biomass accumulations in early stages of benthic communities to the survival of coral recruits. *Proceedings of the Third International Coral Reef Symposium* 1, 15–21.

Blake, W. P. 1907. Lake Cahuilla. *National Geographic Magazine* 2, 62–103.

Boehm, M. C. 1984. An overview of the lithostratigraphy, biostratigraphy, and paleoenvironments of the late Neogene San Felipe marine sequence, Baja California, Mexico. In *Geology of Baja California*. Edited by V. A. Frizzell. San Diego: Pacific Section, Society of Economic Paleontologists and Mineralogists, pp. 219–236.

Bohannon, R. G., and Parsons, T. 1995. Tectonic implications of post–30 Ma Pacific and North American relative plate motions. *Geological Society of America Bulletin* 107, 937–959.

Bosence, D. W. J. 1983. The occurrence and ecology of recent rhodoliths: a review. In *Coated Grains*. Edited by T. M. Peryt. Berlin, Springer-Verlag, pp. 225–242.

Bosence, D. W. J., and Pedley, H. M. 1982. Sedimentology and palaeoecology of a Miocene coralline algal biostrome from the Maltese Islands. *Palaeoclimatology, Palaeoecology, Palaeogeography* 38, 9–43.

Brasier, M. D., Green, O. R., Jephcoat, A. P., Kleppe, A. K., Van Kranendonk, M. J., Lindsay, J. F., Steele, A., and Grassineau, N. V. 2002. Questioning the evidence for Earth's oldest fossils. *Nature* 416, 76–81.

Bray, N. A. 1988a. Thermohaline circulation in the Gulf of California. *Journal of Geophysical Research* 93(C2), 4993–5020.

Bray, N. A. 1988b. Water mass formation in the Gulf of California. *Journal of Geophysical Research* 93(C8), 9223–9240.

Bray, N. A., and Robles, J. M. 1991. Physical oceanography of the Gulf of California. In *The Gulf and Peninsular Province of the Californias*. Edited by J. P. Dauphin and B. R. T. Simoneit. American Association of Petroleum Geologists Memoir 47, pp. 511–553.

Bromley, R. G., and D'Alessandro, A. 1984. The ichnogenus *Entobia* from the Miocene, Pliocene, and Pleistocene of southern Italy. *Rivista Italiana di Paleontologia e Stratigraphia* 90, 227–296.

Brooke, B. 2001. The distribution of carbonate eolianite. *Earth-Science Reviews* 55, 135–164.

Brusca, R. C. 1980. *Common Intertidal Invertebrates of the Gulf of California*. Tucson, University of Arizona Press. 880 p.

Brusca, R. C., and Bryner, G. C. 2004. A case study of two Mexican biosphere reserves: the Upper Gulf of California/Colorado River Delta and Pinacate/Gran Desierto de Altar Biosphere Reserves. In *Science and Politics in the International Environment*. Edited by N. E. Harrison and G. C. Bryner. New York, Rowman and Littlefield, pp. 28–64.

Brusca, R. C., and Thomson, D. A. 1975. Pulmo reef: the only "coral reef" in the Gulf of California. *Ciencias Marinas* 2, 37–53.

Brusca, R. C., Findley, L. T., Hastings, P. A., Hendrickx, M. E., Torre-Cosio, J., and Van der Heiden, A. M. 2005. Macrofaunal diversity in the Gulf of California. In *Biodiversity, Ecosystems, and Conservation in Northern Mexico*. Edited by J. L. E. Cartron, G. Ceballos, and R. S. Felger. Oxford: Oxford University Press, pp. 179–203.

Bryant, D., Burke, L., McManus, J., and Spalding, M. 1998. *Reefs at Risk: A Map-Based Indicator of Threats to the World's Corals Reefs*. Washington, D.C.: WRI/ICLARM/WCMC/UNEP

Budd, A. F. 1989. Biogeography of Neogene Caribbean reef corals and its implications for the ancestry of eastern Pacific reef corals. *Memoir of the Association of Australasian Palaeontologists* 8, 219–230.

Budd, A. F. 1991. Neogene paleontology in the northern Dominican Republic. 11. The family Faviidae (Anthozoa: Scleractinia). Part I. The genera *Montastraea* and *Solenastrea*. *Bulletins of American Paleontology* 101, 1–83.

Budd, A. F. 2000. Diversity and extinction in the Cenozoic history of the Caribbean reefs. *Coral Reefs* 19, 25–35.

Budd, A. F., and Coates, A. G. 1992. Non-progressive evolution in a clade of Cretaceous *Montastraea*-like corals. *Paleobioloy* 18, 425–446.

Budd, A. F., and Johnson, K. G. 1997. Coral reef community dynamics over 8 million years of evolutionary time: stasis and turnover. *Proceedings of the Eighth International Coral Reef Symposium*, Balboa, Panama, pp. 423–428.

Budd, A. F., and Johnson, K. G. 1999a. Neogene paleontology in the northern Dominican Republic. 19. The family Faviidae (Anthozoa: Scleractinia). Part II. The genera *Caulastrea, Favia, Diploria, Thysanus, Hadrophyllia, Manicina*, and *Colpophyllia*. *Bulletins of American Paleontology* 356, 1–83.

Budd, A. F., and Johnson, K. G. 1999b. Origination preceding extinction during late Cenozoic turnover of Caribbean reefs. *Paleobiology* 25, 188–200.

Budd, A. F., Stemann, T. A., and Johnson, K. G. 1994. Stratigraphic distributions of genera and species of Neogene to Recent Caribbean reef corals. *Journal of Paleontology* 68, 951–977.

Budd, A. F., Johnson, K. G., and Stemann, T. A. 1996. Plio-Pleistocene turnover and extinctions in the Caribbean reef-coral fauna. In *Evolution and Environment in Tropical America*. Edited by J. B. C. Jackson, A. F. Budd, and A. G. Coates. Chicago, University of Chicago Press, pp. 168–204.

Buising, A. V. 1990. The Bouse Formation and bracketing units, southeastern California and western Arizona: implications for the evolution of the proto-Gulf of California and the lower Colorado River. *Journal of Geophysical Research* 95(B12) 20.111–20.132.

Cairns, S. D. 1989. Asexual reproduction in solitary Scleractinia. *Proceedings of the Sixth International Coral Reef Symposium* 2, 641–646.

Cairns, S. D. 1991. A revision of the ahermatypic Scleractinia of the Galápagos and Cocos Islands. *Smithsonian Contribution to Zoology* 504, 1–30.

Cairns, S. D. 1994. Scleractinia of the temperate North Pacific. *Smithsonian Contributions to Zoology* 557, 1–150.

Cajal-Medrano, R., Millán-Núñez, R., and Santamaría-del-Angel, E. 1992. Photosynthetic quotients in the Gulf of California during Autumn 1987 in the Central Region and Spring 1989 in Puerto Don Juan. *Ciencias Marinas* 18(3), 1–16.

Calderón-Aguilera, L. E., Reyes-Bonilla, H., and Carriquiry-Beltran, J. D. 2007. El papel de los arrecifes coralinos en el flujo de carbono en el océano: estudios en el Pacífico mexicano. In *Carbono en Ecosistemas Acuáticos de México*. Edited by B. Hernández-de la Torre and G. Gaxiola-Castro. México, D.F., Instituto Nacional de Ecología-SEMARNAT-CICESE, pp. 215–226.

Calvert, S. E. 1966. Factors affecting distribution of laminated diatomaceous sediments in the Gulf of California. In *Marine Geology of the Gulf of California: A Symposium*. Edited by T. H. Van Andel and G. G. Shor Jr. American Assocation of Petroleum Geologists Memoir 3, pp. 311–330.

Canet, C., Prol-Ledesma, R. M., Melgarejo, J. C., and Reyes, A. 2003. Methane-related carbonates formed at submarine hydrothermal springs: a new setting for microbially derived carbonates? *Marine Geology* 199, 245–261.

Canet, C., Prol-Ledesma, R. M., Proenza, J. A., Rubio-Ramos, M. A., Forrest, M. J., Torres-Vera, M. A., and Rodriguez-Diaz, A. A. 2005a. Mn-Ba-Hg mineralization at shallow submarine hydrothermal vents in Bahía Concepción, Baja California Sur, Mexico. *Chemical Geology* 224, 96–112.

Canet, C., Prol-Ledesma, R. M., Torres-Alvarado, I., Gilg, H. A., Villanueva, R. E., and Lozano-Santa Cruz, R. 2005b. Silica-carbonate stromatolites related to coastal hydrothermal venting in Bahía Concepción, Baja California Sur, Mexico. *Sedimentary Geology* 174, 97–113.

Cannon, R. 1966. *The Sea of Cortez*. San Francisco, Sunset. 284 p.

Carbajal, N., Sousa, A. Y., and Durazo, R. 1997. A numerical model of the ex-ROFI of the Colorado River. *Journal of Marine Systems* 12, 17–33.

Cariño-Olivera, M. M. 1995. La pesca y el cultivo de perlas en la región de La Paz (1870–1940). Siglo XIX, septiembre–diciembre. Universidad Autónoma de Nuevo León, Instituto Mora, pp. 27–48.

Cariño-Olivera, M. M. 2000. *Historia de las Relaciones Hombre: Naturaleza en Baja California Sur 1500–1940*. 2nd ed. La Paz, UABCS. 228 p.

Cariño-Olivera, M. M., and Caceres-Martinez, C. 1990. La perlicultura en la península de Baja California a principios de siglo. *Serie Científica, Univesidad Autonoma de Baja California Sur, México* 1, 1–6.

Cariño-Olivera, M. M., and Monteforte, M. 1999. El primer emporio perlero sustentable del mundo. *La Compañía Criadora de Concha y Perla de la Baja California S.A., y sus perspectivas para Baja California Sur*. La Paz, UABCS-FONCA-CONACULTA, 325 p.

Carney, R. S. 1994. Consideration of the oasis analogy for chemosynthetic communities at Gulf of Mexico hydrocarbon vents. *Geo-Marine Letters* 14, 149–159.

Carranza-Edwards, A., Bocanegra-García, G., Rosales-Hoz, L., and Galán, L. P. 1998. Beach sands from Baja California peninsula, Mexico. *Sedimentary Geology* 119, 263–274.

Carreño, A. L. 1981. Ostrácodos y foraminíferos planctónicos de la loma del Tirabuzón, Santa Rosalía, Baja California Sur, e implicaciones bioestratigráficas y paleoecológicas. *Universidad Nacional Autónoma de México, Instituto de Geología, Revista* 5(1), 55–64.

Carreño, A. L. 1992. Neogene microfossils from the Santiago Diatomite, Baja California Sur, Mexico. *Paleontología Mexicana,Instituto de Geología, Universidad Nacional Autónoma de México* 59, 1–37

Carreño, A. L., and Helenes, J. 2002. Geology and ages of the islands. In *A New Island Biogeography of the Sea of Cortés*. Edited by T. J. Case, M. L. Cody, and E. Equarra. Oxford: Oxford University Press, pp. 14–40.

Carreño, A. L., and Smith, J. T. 2007. Stratigraphy and correlation for the ancient Gulf of California peninsula, Mexico. *Bulletin of American Paleontology* 371, 1–146.

Carrillo, G. Y., Shaw, W. W., and De Steiguer, J. E. 2001. Non-regulatory restoration of wetlands in the Colorado River delta, México. M.S. thesis, University of Arizona. 135 p.

Carriquiry, J. D., and Sánchez, A. 1999. Sedimentation in the Colorado River delta and upper Gulf of California after nearly a century of discharge loss. *Marine Geology* 158, 125–145.

Carter, R. W. G., and Wilson, P. 1990. The geomorphological, ecological and pedological development of coastal foredunes at Magilligan Point, Northern Ireland. In *Coastal Dunes: Form and Process*. Edited by K. F. Nordstrom, N. P. Psuty, and W. R. G. Carter. New York, John Wiley and Sons, pp. 129–157.

Castro R., Lavín, M. F., and Ripa, P. 1994. Seasonal heat balance in the Gulf of California. *Journal of Geophysical Research* 99(C2), 3249–3261.

Castro, R., Mascarenhas, A. S., Durazo, R., and Collins, C. A. 2000. Seasonal variation of the temperature and salinity at the entrance to the Gulf of California. *Ciencias Marinas* 26, 561–583.

Chave, K. E., Smith, E. S., and Roy, K. J. 1972. Carbonate production by coral reefs. *Marine Geology* 12, 123–140.

Chávez-Romo, H. E., and Reyes-Bonilla, H. 2007. Sexual reproduction of the coral *Pocillopora damicornis* in the southern Gulf of California, México. *Ciencias Marinas* 33, 495–501.

Cintra-Buenrostro, C. E., Foster, M. S., and Meldahl, K. H. 2002. Response of nearshore marine assemblages to global change: a comparison of molluscan assemblages in Pleistocene and modern rhodolith beds in the southwestern Gulf of California, Mexico. *Palaeogeography, Palaeoclimatology, Palaeoecolgy* 183, 229–230.

Cintra-Buenrostro, C., Avila-Serrano, G., and Flessa, W. K. 2005. Who cares about a vanishing clam? Trophic importance of *Mulinia coloradoensis* inferred from predatory damage. *Palaios* 20, 296–302.

Cisneros-Mata, M. A., Montemayor-López, G., and Román-

Rodriguez, M. J. 1995. Life history and conservation of *Totoaba macdonaldi. Conservation Biology* 9(4), 806–814.

Clark, K. R., and Warwick, R. M. 2001. *Change in Marine Communities: An Approach to Statistical Analysis and Interpretation.* London, Primer-E. 320 p.

Clark, R. N. 2000. The chiton fauna of the Gulf of California rhodolith beds (with descriptions of four new species). *Nemouria: Occasional Papers of the Delaware Museum of Natural History* 43, 1–20.

Clarke, W. B., Beg, M. A., and Craig, H. 1969. Excess He-3 in the sea: evidence for terrestrial primordial helium. *Earth and Planetary Science Letters* 6, 213–220.

Cocito, S., Bianchi, C. N., Morri, C., and Peirano, A. 2000. First survey of sessile communities on subtidal rocks in an area with hydrothermal vents. *Hydrobiologia* 426, 113–121.

Cohen, M. J., and Henges-Jeck, C. 2001. *Missing Water: The Uses and Flows of Water in the Colorado River Delta Region.* Oakland, Calif., Pacific Institute for Studies in Development, Environment, and Security. 44 p.

Colgan, M. W. 1990. El Niño and the history of Eastern Pacific reef building. In *Global Ecological Consequences of the 1982–83 El Niño-Southern Oscillation.* Edited by P. W. Glynn. Amsterdam, Elsevier, pp. 183–232.

Colley, S. B., Feingold, J. S., and Glynn, P. W. 2002. Reproductive ecology of *Diaseris distorta* (Michelin) (Fungiidae) in the Galápagos Islands. *Proceedings of the Ninth International Coral Reef Symposium* 1, 373–379.

Conly, A. G., Beaudoin, G., and Scott, S. D. 2006. Isotopic constraints on fluid evolution and precipitation mechanisms for the Boléo Cu-Co-Zn deposit, Mexico. *Mineralium Deposita* 41, 127–151.

Conway, N. M., Kennicutt II, M. C., and Van Dover, C. L. 1994. Stable isotopes in the study of marine chemosynthetic-based ecosystems. In *Stable Isotopes in Ecology and Environmental Science.* Edited by K. Lajtha and R. H. Michener. London, Blackwell Science, pp. 158–186.

Cortés, J. 1986. Biogeografía de corales hermatípicos: el istmo Centro Americano. *Anales del Instituto de Ciencias del Mar y Limnología* 13, 297–304.

Cortés, J., ed. 2003. *Latin American Coral Reefs.* Amsterdam, Elsevier. 497 p.

Cruz-Ayala, M. B., Núñez-López, R. A., and López, G. E. 2001. Seaweeds in the southern Gulf of California. *Botanica Marina* 44, 187–197.

Damon, P. E., Shafiqullach, M., and Scarborough, R. B. 1978. The evolution of the lower Colorado River revised chronology for critical stages. *Geological Society of America Abstracts with Programs* 10(3), 101–102.

Dana, T. F. 1975. Development of contemporary Eastern Pacific corals reefs. *Marine Biology* 33, 355–374.

Dawson, E. Y. 1944. The marine algae of the Gulf of California. *Allan Hancock Pacific Expeditions* 3, 1–452.

Dawson, E. Y. 1960. Marine red algae of Pacific México. Part III. Cryptonemiales, Corallinaceae. *Pacific Naturalist* 2, 3–125.

Dean, M. A. 1996. Neogene Fish Creek gypsum and associated stratigraphy and paleontology, southwestern Salton Trough, California. In *Sturzstroms and Detachment Faults, Anza-Borrego Desert State Park, California.* Edited by P. Abbott and D. Seymour. South Coast Geological Society Annual Field Trip Guide Book 24, pp. 123–148.

DeDiego-Forbis, T., Douglas, R., Gorsline, D. S., Nava-Sanchez, E. H., Mack, L., and Banner, J. 2004. Late Pleistocene (Last Interglacial) terrace deposits, Bahía Coyote, Baja California Sur, México. *Quaternary International* 120, 29–40.

De Grave, S., and Whitaker, A. 1999. Benthic community readjustment following dredging of a muddy-maerl matrix. *Marine Pollution Bulletin* 38, 102–108.

De Grave, S., Fazakerley, H., Kelly, L., Guiry, M. D., Ryan, M., and Walshe, J. 2000. *A Study of Selected Maërl Beds in Irish Waters and Their Potential for Sustainable Extraction.* Marine Resource Series 10, 44 p.

De Mets, C. 1995. A reappraisal of seafloor spreading lineations in the Gulf of California: implications for the transfer of Baja California to the Pacific plate and estimates of Pacific–North America motion. *Geophysical Research Letters* 22, 3545–3548.

Díaz-Uribe, J. G., Arreguín-Sánchez, F., and Cisneros-Mata, M. A. 2007. Multispecies perspective for small-scale fisheries management: a trophic analysis of La Paz Bay in the Gulf of California, Mexico. *Ecological Modeling* 201, 205–222.

Dibblee, T. W., Jr. 1996. Stratigraphy and tectonics of the Vallecitos–Fish Creek Mountains, Vallecito Badlands, Coyote Mountains, and Yuha Desert, southwest Imperial Basin. In *Sturzstroms and Detachment Faults, Anza Borrego Desert State Park, California.* Edited by P. L. Abbott and D. C. Seymor. South Coast Geological Society Field Guidebook 24, pp. 59–79.

Diguet, L. 1895. Note sur une exploration de la Basse-California par M. Diguet, chargé dune mission par le Museum. *Bulletin du Museum D'Histoire Naturelle* 1, 29–31.

Diguet, L. 1899. Etude dur l'exploitation de l'huître perlière dans el golfe de Californie. *Bulletin de la Societé Centrale d'Aquiculture et de Pêche* 7, 1–15.

Diguet, L. 1911. Pêcheries du Golfe de Californie. Poissons, cétacés, phoques, loutres, perles, culture de la nacre et repartition géographique des gisementes perliers. *Bulletin de la Societé Centrale d'Aquiculture et de Pêche* 8, 186–196.

Dorsey, R. J. 1997. Origin and significance of rhodolith-rich strata in the Punta El Bajo section, southeastern Pliocene Loreto basin. In *Pliocene Carbonates and Related Facies Flanking the Gulf of California, Baja California, Mexico.* Edited by M. E. Johnson and J. Ledesma-Vázquez. Geological Society of America Special Paper 318, pp. 119–126.

Dorsey, R. J., and Kidwell, S. M. 1999. Mixed carbonate-siliciclastic sedimentation on a tectonically active margin: example from the Pliocene of Baja California Sur, Mexico. *Geology* 27, 935–938.

Dorsey, R. J., Stone, K. A., and Umhoefer, P. J. 1997. Stratigraphy, sedimentology, and tectonic development of the southeastern Pliocene Loreto Basin, Baja California Sur, Mexico. In *Pliocene Carbonates and Related Facies Flanking the Gulf of California, Baja California, Mexico.* Edited by M. E. Johnson and J. Ledesma-Vázquez. Geological Society of America Special Paper 318, pp. 83–109.

Dorsey, R. J., Umhoefer, P. J., Ingle, J. C., and Mayer, L. 2001. Late Miocene to Pliocene stratigraphic evolution of northeast Isla Carmen, Gulf of California: implications for oblique-rifting tectonics. *Sedimentary Geology* 144, 97–123.

Douglas, M. W., Maddox, R. A., and Howard, K. 1993. The Mexican monsoon. *Journal of Climate* 6, 1665–1677.

Douglas, R., Gonzalez-Yajimovich, O., Ledesma-Vázquez, J., and Staines-Urias, F. 2007. Climate forcing, primary production and the distribution of Holocene biogenic sediment in the Gulf of California. *Quaternary Science Reviews* 26, 115–129.

Durham, J. W. 1947. *Corals from the Gulf of California and the North Pacific Coast of America.* Geological Society of America Memoir 20. 68 p.

Durham, J. W. 1950. Megascopic paleontology and marine stratigraphy. In *1940 E. W. Scripps Cruise to the Gulf of California.* Part 2. By C. A. Anderson et al. Geological Society of America Memoir 43. 216 p.

Durham, J. W. 1966. Coelenterates, especially stony corals, from the Galápagos and Cocos Islands. In *The Galápagos.* Edited by R. L. Bowman. Berkeley, University of California Press, pp. 123–135.

Durham, J. W. 1980. A new fossil *Pocillopora* (coral) from Guadalupe Island, México. In *The California Islands.* Edited by D. M. Powers. Santa Barbara, Natural History Museum of Santa Barbara, pp. 63–70.

Durham, J. W., and Allison, E. C. 1960. The geologic history of Baja California and its marine faunas. *Systematic Zoology* 9, 47–91.

Durham, J. W., and Barnard, J. L. 1952. Stony corals of the eastern Pacific collected by the Velero III and Velero IV. *Allan Hancock Pacific Expedition* 16, 1–110.

Eberly, L. D., and Stanley, T. B., Jr. 1978. Cenozoic stratigraphy and geologic history of southwestern Arizona. *Geological Society of America Bulletin* 89, 921–940.

Edinger, E. N., Pandolfi, J. M., and Kelley, R. A. 2001. Community structure of Quaternary coral reefs compared with Recent life and death assemblages. *Paleobiology* 27, 669–694.

Edmond, J. M., Von Damm, K. L., McDuff, R. E., and Measures, C. I. 1982. Chemistry of hot springs on the East Pacific Rise and their effluent dispersal. *Nature* 297, 187–191.

Enríquez-Andrade R., Anaya-Reyna, G., Barrera-Guevara, J. C., Carvajal-Moreno, M. A., Martínez-Delgado, M. E., Vaca-Rodríguez, J., and Valdés-Casillas, C. 2005. An analysis of critical areas for biodiversity conservation in the Gulf of California. *Ocean and Coastal Management* 48, 31–50.

Eros, J. M., Johnson, M. E., and Backus, D. H. 2006. Rocky shores and development of the Pliocene-Pleistocene Arroyo Blanco basin on Isla Carmen in the Gulf of California, Mexico. *Canadian Journal of Earth Sciences* 43, 1149–1164.

Escalona-Alcázar, F. J., Delgado-Argote, L. A., López-Martínez, M., and Rendón-Márquez, G. 2001. Late Miocene volcanism and marine incursions in the San Lorenzo Archipelago, Gulf of California, Mexico. *Revista Mexicana de Ciencias Geológicas* 18(2), 111–128.

Fenby, S. S., and Gastil, R. G. 1989. Geologic-tectonic map, Gulf and Peninsular Province of the Californias. In *The Gulf and Peninsular Province of the Californias.* Edited by J. P. Dauphin and B. R. T. Simoneit. American Association of Petroleum Geologists Memoir 47, map plates 9 and 10.

Fisher, A. T., Giambalvo, E., Sclater, J., Kastner, M., Ransom, B., Weinstein, Y., and Lonsdale, P. 2001. Heat flow, sediment and pore fluid chemistry, and hydrothermal circulation on the east flank of Alarcon Ridge, Gulf of California. *Earth and Planetary Science Letters* 188, 521–534.

Flessa, K. W., and Téllez, M. A. 2001. *Taxonomic Status and Distribution of the Bivalve Mollusk Mulinia coloradoensis in the Gulf of California.* Report to the Center of Biological Diversity and Defenders of the Wildlife. 11 p.

Fletcher, J. M., Kohn, B. P., Foster, D. A., and Gleadow, A. J. W. 2000.

Heterogeneous Neogene cooling and exhumation of the Los Cabos block, Southern Baja California: evidence from fission-track thermochronology. *Geology* 28, 107–110.

Fletcher, J. M., Pérez-Venzor, J. A., González-Barba, G., and Aranda-Gómez, J. J. 2003. Ridge-trench interactions and the ongoing capture of the Baja California Microplate: new insights from the Southern Gulf Extensional Province. In *Geological Transects across Cordilleran Mexico.* Guidebook for Field Trips of the 99th Annual Meeting of the Cordilleran Section, GSA, Mexico, D.F., UNAM, IG, Publicación Especial 1, 2, pp. 13–31.

Forrest, M. J. 2004. The geology, geochemistry, and ecology of a shallow water submarine hydrothermal vent in Bahía Concepción, Baja California Sur, Mexico. M.S. thesis, Moss Landing Marine Labs, Moss Landing, Calif. 112 p.

Forrest, M. J., Ledesma-Vázquez, J., Ussler, W., Kulongoski, J. T., Hilton, D. R., and Greene, H. G. 2005. Gas geochemistry of a shallow submarine hydrothermal vent associated with El Requesón fault zone in Bahía Concepción, Baja California Sur, México: *Chemical Geology* 224, 82–95.

Forsman, Z. 2003. Phylogeny and phylogeography of *Porites* and *Siderastrea* (Scleractinia: Cnidaria) species in the Caribbean and eastern Pacific, based on the nuclear ribosomal ITS region. Ph.D. diss., Houston, University of Houston. 153 p.

Fossa, J. H., Mortensen, P. B., and Furevic, D. M. 2002. The deepwater coral *Lophelia pertusa* in Norwegian waters: distribution and fishery impacts. *Hydrobiologia* 417, 1–12.

Foster, A. B. 1979. Environmental variation in a fossil scleractinian coral. *Lethaia* 12, 245–264.

Foster, A. B. 1980. Ecology and morphology of the Caribbean Mio-Pliocene reef-coral *Siderastrea*. *Acta Palaeontologica Polonica* 25, 439–450.

Foster, M. S. 2001. Rhodoliths: between rock and soft places. *Journal of Phycology* 37, 659–667.

Foster, M. S., Riosmena-Rodríguez, R., Steller, D. L., and Woelkerling, W. J. 1997. Living rhodolith beds in the Gulf of California and their implications for paleoenvironmental interpretation. In *Pliocene Carbonates and Related Facies Flanking the Gulf of California, Baja California, Mexico.* Edited by M. E. Johnson and J. Ledesma-Vázquez. Geological Society of America Special Paper 318, pp. 127–139.

Foster, M. S., McConnico, L. M., Lundsten, L., Wadsworth, T., Kimball, T., Brooks, L. B., Medina-Lopez, M. A., Riosmena-Rodríguez, R., Hernandez-Carmona, G., Vázquez-Slizondo, R. M., Johnson, S., and Steller, D. L. 2007. The diversity and natural history of a *Lithothamnion muelleri–Sargassum horridum* community in the Gulf of California. *Ciencias Marinas* 33, 367–384.

Frost, S. H. 1977a. Miocene to Holocene evolution of Caribbean province reef-building corals. *Proceedings of the Third International Coral Reef Symposium,* University of Miami, Miami, pp. 353–359.

Frost, S. H. 1977b. Oligocene reef coral biogeography Caribbean and western Tethys. *Mémories du Bureau Recherches Géologiques et Minières* 89, 342–352.

Fukami, H., Budd, A. F., Paulay, G., Solé-Cava, A., Chen, C. A., Iwao, K., and Knowlton, N. 2004. Conventional taxonomy obscures deep divergence between Pacific and Atlantic corals. *Nature* 427, 832–835.

Galindo-Bect, S. M. 2003. Larvas y postlarvas de camarones peneidos en el Alto Golfo de California y capturas de camarón con relación al flujo del Rio Colorado. Ph.D. diss., Facultad de Ciencias Marinas, Universidad Autónoma de Baja California, Ensenada, B.C. 160 p.

Gastil, R. G., and Krummenacher, D. 1977. Reconnaissance geology of coastal Sonora between Puerto Lobos and Bahia Kino. *Geological Society America Bulletin* 88, 189–198.

Gastil, R. G., Phillips, R. P., and Allison, E. C. 1973. Reconnaissance geologic map of the State of Baja California. Geological Society of America, 3 map sheets at scale of 1:250,000.

Gastil, R. G., Krummenacher, D., and students at San Diego State University. 1974. Reconnaissance geology of coastal Sonora between Puerto Lobos and Bahia Kino. Geological Society America, 1 map sheet at scale of 1:150,000.

Gastil, R. G., Phillips, R. P., and Allison, E. C. 1975. *Reconnaissance Geology of the State of Baja California*. Geological Society of America Memoir 140. 170 p.

Gastil, R. G., Krummenacher, D., and Minch, J. 1979. The record of Cenozoic volcanism around the Gulf of California. *Geological Society of America Bulletin* 90, 839–857.

Gastil, R. G., Minch, J., and Phillips, R. P. 1983. The geology and ages of islands. In *Island Biogeography in the Sea of Cortés*. Edited by T. J. Case and M. L. Cody. Berkeley, University of California Press, pp. 13–15.

Gastil, R. G., Neuhaus, J., Cassidy, M., Smith, J. T., Ingle, J. C., and Krummenacher, D. 1999. Geology and paleontology of southwestern Isla Tiburón, Sonora, México. *Revista Mexicana de Ciencias Geológicas* 16, 1–34.

Geister, J. 1975. Ocurrence of Pocillopora in late Pleistocene Caribbean coral reefs. In *Second Symposium International sur les Coraux et Récifs Coralliens Fossiles*. Mémories du Bureau Recherches Géologiques et Minières, pp. 378–388.

Gendrop-Funes, V., Acosta-Ruíz, M. J., and Schwartzlose, R. A. 1978. Distribución horizontal de la clorofila "a" durante la primavera en la parte Norte del Golfo de California. *Ciencias Marinas* 5, 71–89.

German, C. R., Baker, E. T., and Klinkhammer, G. 1995. Regional setting of hydrothermal activity. *Geological Society of London, Special Publications* 87, 3–15.

Giggenbach, W. F. 1997. Relative importance of thermodynamic and kinetic processes in governing the chemical and isotopic composition of carbon gases in high-heatflow sedimentary basins. *Geochimica et Cosmochimica Acta* 61(17), 3763–3785.

Gilbert, J. Y., and Allen, W. A. 1943. The phytoplankton of the Gulf of California obtained by the E. W. Scripps in 1939 and 1940. *Journal of Marine Resources* 5, 89–110.

Glynn, P. W. 1988. El Nino–Southern Oscillation 1982–1983: near shore population, community, and ecosystem responses. *Annual Review of Ecology and Systematics* 19, 309–345.

Glynn, P. W. 2001. Eastern Pacific coral reef ecosystems. In *Coastal Marine Ecosystems of Latin America*. Edited by U. Seeliger and B. Kjerfve. Berlin, Springer, pp. 281–305.

Glynn, P. W., and Ault, J. S. 2000. A biogeographic analysis and review of the far eastern Pacific coral reef region. *Coral Reefs* 19, 1–23.

Glynn, P. W., and Wellington, G. M. 1983. *Corals and Coral Reefs of the Galápagos Islands*. Berkeley, University of California Press. 330 p.

Glynn, P. W., Gassman, N. J., Eakin, C. M., Cortés, J., Smith, D. B., and Guzmán, H. M. 1991. Reef coral reproduction in the eastern Pacific: Costa Rica, Panamá, and Galápagos Islands (Ecuador). I. Pocilloporidae. *Marine Biology* 109, 355–368.

Glynn, P. W., Colley, S. B., Eakin, C. M., Smith, D. B., Cortés, J., Gassman, N. J., Guzmán, H. M., Del Rosario, J. B., and Feingold, J. S. 1994. Reef coral reproduction in the eastern Pacific: Costa Rica, Panamá, and Galápagos Islands (Ecuador). II. Poritidae. *Marine Biology* 118, 191–208.

Glynn, P. W., Colley, S. B., Gassman, N. J., Black, K., Cortés, J., and Maté, J. L. 1996. Reef coral reproduction in the eastern Pacific: Costa Rica, Panamá, and Galápagos Islands (Ecuador). III. Agariciidae (*Pavona gigantea* and *Gardineroseris planulata*). *Marine Biology* 125, 579–601.

Goodwin, D. H., Flessa, K. W., Schöme, B. R., and Dettman, D. L. 2001. Cross-calibration of daily increments, stable isotope variation and temperature in the Gulf of California bivalve mollusk *Chione cortezi*: implication for paleoenvironmental analysis. *Palaios* 16, 378–398.

Grall, J., and Glemarec, M. 1997. Biodiversite des fonds de maerl en Bretagne: approache fonctionnelle et impacts anthropogeniques. *Vie Milieu* 47, 339–349.

Grall, J., Le Loc'h, F., Guyonnet, B., and Riera, P. 2006. Community structure and food web based on stable isotopes (Δ^{15}N and Δ^{13}C) analysis of a North Eastern Atlantic maerl bed. *Marine Ecology Progress Series* 338, 1–15.

Grewingk, C. 1848. Beitrag zur Kenntniss der geognostischen Beschaffenheit Californiens. *Verhandl Russisch Min Gesell St. Petersburg* 1847, 142–162.

Gutiérrez-Estrada, M., and Ortiz-Perez, M. A. 1985. Pacific Mexico. In *The World's Coastline*. Edited by E. C. F. Bird and M. L. Schwartz. New York, Van Nostrand Reinhold, pp. 37–40.

Halfar, J., and Mutti, M. 2005. Global dominance of coralline red-algal facies: a response to Miocene oceanographic events. *Geology* 33, 481–484.

Halfar, J., Kronz, A., and Zachos, J. C. 2000. Growth and high-resolution paleoenvironmental signals of rhodoliths (coralline red algae): a new biogenic archive. *Journal of Geophysical Research: Oceans* 105, 22107–22116.

Halfar, J., Godinez-Orta, L., Goodfriend, G. A., Mucciarone, D. A., Ingle, J. C., and Holden, P. 2001. Holocene-Pleistocene carbonate sedimentation and tectonic history of the La Paz area, Baja California Sur, Mexico. *Sedimentary Geology* 144, 149–177.

Halfar, J., Godinez-Orta, L., Mutti, M., Valdez-Holguin, J. E., and Borges, J. M. 2004. Nutrient and temperature controls on modern carbonate production: an example from the Gulf of California, Mexico. *Geology* 32, 213–216.

Hall, H. 1992. Shadows in a Desert Sea. Film for PBS produced for *Nature* and the BBC by Howard Hall Productions. http://www .howardhall.com/tv/shadows.html

Hall-Spencer, J. M. 1998. Conservation issues relating to maerl beds as habitats for mollusks. *Journal of Conchology, Special Publication* 2, 271–286.

Hall-Spencer, J. M. 1999. Effects of towed demersal fishing gear on biogenic sediments: a 5-year study. In *Impact of Trawl Fishing on Benthic Communities*. Edited by O. Giovanardi. Rome, ICRAM, pp. 9–20.

Hall-Spencer, J. M., and Moore, P. G. 2000. Impact of scallop dredging on maerl grounds. In *Effects of Fishing on Nontarget Species and Habitats: Biological, Conservation and Socioeconomic issues*. Edited by M. J. Kaiser and S. J. de Groot. Oxford, Blackwell Science, pp. 105–117.

Hall-Spencer, J. M., Grall, J., Moore, P. G., and Atkinson, R. J. A. 2003. Bivalve fishing and maerl-bed conservation in France and the UK: retrospect and prospect. *Aquatic Conservation Marine and Freshwater Ecosystems* 13, S33–S41.

Hall-Spencer, J. M., White, N., Gillespie, E., Gillham, K., and Foggo, A. 2006. Impact of fish farms on maerl beds in strongly tidal areas. *Marine Ecology Progress Series* 326, 1–9.

Hall-Spencer, J. M., Rodolfo-Metalpa, R., Martin, S., Ransome, E., Fine, M., Turner, S. M., Rowley, S. J., Tedesco, D., Buia, M.-C. 2008. Volcanic carbon dioxide vents show ecosystem effects of ocean acidification. *Nature* 454, 96–99.

Hanna, G. D., and Hertlein, L. G. 1927. Expedition of the California Academy of Sciences to the Gulf of California in 1921. *Proceedings of the California Academy of Sciences* 16, 137–157.

Haq, B. U., Hardenbol, J., and Vail, P. R. 1987. Chronology of fluctuating sea levels since the Triassic. *Science* 235, 1156–1167.

Hariot, P. 1895. Algues du Golfe de California recuellies par M. Diguet. *Journal Botanic* 9, 167–170.

Harris, P. T., Tsuji, Y., Marshall, J. F., Davies, P. J., Honda, N., and Matsuda, H. 1996. Sand and rhodolith-gravel entrainment on the mid- to outer-shelf under a western boundary current: Fraser Island continental shelf, eastern Australia. *Marine Geology* 129, 313–330.

Hastings, J. R., and Turner, R. M. 1965. Seasonal precipitation regimes in Baja California, Mexico. *Geografiska Annaler* 47A, 204–223.

Hausback, B. P. 1984. Cenozoic volcanic and tectonic evolution of Baja California Sur, Mexico. In *Geology of the Baja California.* Vol. 39. Edited by V. A. Frizzell. Pacific Section, Society of Economic Paleontologists and Mineralogists, pp. 219–236.

Hayes, M. L., Johnson, M. E., and Fox, W. T. 1993. Rocky-shore biotic associations and their fossilization potential: Isla Requeson (Baja California Sur, Mexico). *Journal of Coastal Research* 9, 944–957.

Heck, K. L., Jr., and McCoy, E. D. 1978. Long-distance dispersal and the reef-building corals of the Eastern Pacific. *Marine Biology* 48, 349–356.

Helenes, J., and Carreño, A. L. 1999. Neogene sedimentary evolution of Baja California in relation to regional tectonics. In *Earth Sciences in Mexico: Some Recent Perspectives.* Edited by C. Lomnitz. Special Issue of *Journal of South American Sciences* 12, 589–605.

Helenes, J., Carreño, A. L., Esparza, M. A., and Carrillo, R. M. 2005. Neogene paleontology in the Gulf of California and the geologic evolution of Baja California. *Proceeds of the Seventh International Meeting on the Geology of the Baja California Peninsula,* Ensenada, Baja California, p. 2.

Hendrickx, M. E., Brusca, R. C., and Findley, L. T. 2005. *A Distributional Checklist of the Macrofauna of the Gulf of California, Mexico.* Part 1. *Invertebrates.* Tucson, Arizona-Sonora Desert Museum. 429 p.

Hernández-Ayón, J. M., Galindo-Bect, M. S., Flores-Báez, B. P., and Álvarez-Borrego, S. 1993. Nutrient concentrations are high in the turbid waters of the Colorado River delta. *Estuary Coastal Shelf Science* 37, 593–602.

Hertlein, L. G. 1957. Pliocene and Pleistocene fossils from the southern portion of the Gulf of California. *Bulletin of Southern California Academy of Sciences* 56, 57–75.

Hertlein, L. G. 1966. Pliocene fossils from Ranco El Refugio, Baja California, and Cerralvo Island, Mexico. *Proceedings of the California Academy of Sciences* 30, 256–284.

Hertlein, L. G. 1972. Pliocene fossils from Baltra (South Seymour) Island, Galapagos Islands. *Proceedings of the California Academy of Sciences* 39, 25–46.

Hertlein, L. G., and Emerson, W. K. 1959. Results of the Puritan-American Museum of Natural History Expedition to western México. Part 5. Pliocene and Pleistocene megafossils from Tres Marías Islands. *American Museum Novitates* 1940, 1–15.

Hetzinger, J., Halfar, J., Riegl, B., and Godinez-Orta, L. 2006. Sedimentology and acoustic mapping of modern rhodolith beds on a non-tropical carbonate shelf (Gulf of California, Mexico). *Journal of Sedimentary Research* 76, 670–682.

Hinojosa-Arango, G., and Riosmena-Rodríguez, R. 2004. Influence of rhodolith-forming species and growth form on associated fauna of rhodolith beds in the central west Gulf of California, Mexico. *Pubblicazioni della Stazione Zoologica di Napoli Marine Ecology* 25, 109–127.

Holt, J. W., Holtt, W. E., and Stock, J. M. 2000. An age constraint on Gulf of California rifting from the Santa Rosalía basin, Baja California Sur, Mexico. *Geological Society of America Bulletin* 112, 540–549.

Iglesias-Prieto, R., Reyes-Bonilla, H., and Riosmena-Rodriguez, R. 2003. Effects of 1997–1998 ENSO on coral reef communities in the Gulf of California, Mexico. *Geofísica Internacional* 42, 1–5.

Iglesias-Prieto, R., Beltrán Ramírez, V. H., LaJeunesse, T. C., Reyes-Bonilla, H., and Thomé, P. E. 2004. Different algal symbionts explain the vertical distribution of dominant reef corals in the eastern Pacific. Proceedings of the Royal Society of London, series B 271, 1757–1763.

Iijima, A. 1988. Silica diagenesis. Part II. In Diagenesis II. Edited by G. V. Chilingarian and K. H. Wolf. Developments in Sedimentology 43, pp. 189–211.

Ives, R. I. 1959. Shell dunes of the Sonoran shore. *American Journal of Science* 257, 449–457.

James, D. W. 2000. Diet, movement, and covering behavior of the sea urchin *Toxopneustes roseus* in rhodolith beds in the Gulf of California, Mexico. *Marine Biology* 137, 913–923.

James, D. W., Foster, M. S., and O'Sullivan, J. 2006. Bryoliths (Bryozoa) in the Gulf of California. *Pacific Science* 60, 117–124.

Jenden, P. D., Kaplan, I. R., Poreda, R. J., and Craig, H. 1988. Origin of nitrogen-rich gases in the California Great Valley: evidence from helium, carbon, and nitrogen isotope ratios. *Geochimica et Cosmochimica Acta* 52, 851–861.

Johnson, A. F. 1977. A survey of the strand and dune vegetation along the Pacific and southern coasts of Baja California, Mexico. *Journal of Biogeography* 7, 83–99.

Johnson, A. F. 1982. Dune vegetation along the eastern shore of the Gulf of California. *Journal of Biogeography* 9, 317–330.

Johnson, K. G. 2001. Middle Miocene recovery of Caribbean reef corals: new data from the Tamana Formation, Trinidad. *Journal of Paleontology* 75, 513–526.

Johnson, M. E. 2002a. *Discovering the Geology of Baja California: Six Hikes on the Southern Gulf Coast.* Tucson, University of Arizona Press. 221 p.

Johnson, M. E. 2002b. Ancient islands in the stream: paleogeography and expected circulation patterns. *Geobios* 35, 96–106.

Johnson, M. E., and Hayes, M. L. 1993. Dichotomous facies on a Late Cretaceous rocky island as related to wind and wave patterns (Baja California, Mexico). *Palaios* 8, 385–395.

Johnson, M. E., and Ledesma-Vázquez, J. 1999. Biological zonation on a rocky-shore boulder deposit: Upper Pleistocene Bahía San Antonio (Baja California Sur, Mexico). *Palaios* 14, 569–584.

Johnson, M. E., and Ledesma-Vázquez, J. 2001. Pliocene-Pleistocene rocky shorelines trace coastal development of Bahia Concepcion, gulf coast of Baja California Sur (Mexico). *Palaeogeography, Palaeoclimatology, Palaeoecology* 166, 65–88.

Johnson, M. E., Ledesma-Vázquez, J., Mayall, M. A., and Minch, J. 1997. Upper Pliocene stratigraphy and depositional systems: the Peninsula Concepción basins in Baja California Sur. In *Pliocene Carbonates and Related Facies Flanking the Gulf of California, Baja California, Mexico*. Edited by M. E. Johnson and J. Ledesma-Vázquez. Geological Society of America Special Paper 318, pp. 57–72.

Johnson, M. E., Backus, D. H., and Ledesma-Vázquez, J. 2002. Mass gravity slide from the Upper Pliocene of Mesa Barracas, Baja California Sur, México. *Abstracts of the Sixth International Meeting on Geology of the Baja California Peninsula*, La Paz, pp. 33–34.

Johnson, M. E., Backus, D. H., and Ledesma-Vázquez, J. 2003. Offset of Pliocene ramp facies at El Mangle by El Coloradito fault, Baja California Sur: implications for transtensional tectonics. In *Tectonic Evolution of Northwestern Mexico and the Southwestern USA*. Edited by S. E. Johnson, S. R. Patterson, J. M. Fletcher, G. H. Girty, D. L. Kimbrough, and A. Martin-Barajas. Geological Society of America Special Paper 374, pp. 407–420.

Johnson, M. E., López-Pérez, R. A., Ransom, C. R., and Ledesma-Vázquez, J. 2007. Late Pleistocene coral-reef development on Isla Coronados, Gulf of California. *Ciencias Marinas* 33, 105–120.

Johnson, N. M., Officer, C. B., Opdyke, N. D., Woodward, G. D., Zietler, P. K., and Lindsay, E. H. 1983. Rates of Late Cenozoic tectonism in the Vallecitos–Fish Creek Basin, western Imperial Valley, California. *Geology* 11, 664–667.

Jordan, E. K., and Hertlein, L. G. 1926. Expedition to the Revillagigedo Islands México in 1925. IV. *Proceedings of the California Academy of Sciences* 15, 209–217.

Kamenos, N. A., Moore, P. G., and Hall-Spencer, J. M. 2004. Nursery-area function of maerl grounds for juvenile queen scallops *Aequipecten opercularis* and other invertebrates. *Marine Ecology Progress Series* 274, 183–189.

Karig, D. E., and Jensky, W. 1972. The proto-Gulf of California. *Earth and Planetary Science Letters* 17, 169–174.

Keen, A. M. 1971. *Sea Shells of Tropical West America*. 2nd ed. Stanford, Calif., Stanford University Press. 1064 p.

Keen, A. M., and Coan, E. 1974. *Marine Molluscan Genera of Western North America: An Illustrated Key*. 2nd ed. Stanford, Calif., Stanford University Press. 208 p.

Knox, G. A. 2001. *The Ecology of Seashores*. Boca Raton, Fla.: CRC Press. 557 p.

Kocurek, G. A. 1996. Desert Aeolian systems. In *Sedimentary Environments: Processes, Facies and Stratigraphy*. 3rd ed. Edited by H. G. Reading. Oxford, Blackwell Science, pp. 125–153.

Konhauser, K. O., Jones, B., Reysenbach, A. L., and Renaut, R. W. 2003. Hot spring sinters: keys to understanding Earth's earliest life forms. *Canadian Journal of Earth Sciences* 40, 1713–1724.

Kowalewski, M. 1995. Quantitative taphonomy, ecology, and paleoecology of shelly invertebrates from the intertidal environments of the Colorado River delta, northeastern Baja California, Mexico. Ph.D. diss., University of Arizona, Tucson. 348 p.

Kowalewski, M., Flessa, K. A., and Aggen, J. A. 1994. Taphofacies analysis of recent shelly cheniers (beach ridges) northeastern Baja California, Mexico. *Facies* 31, 209–242.

Kowalewski, M., Goodfriend, G. A., and Flessa, K. W. 1998. High-resolution estimates of temporal mixing within shell beds: the evils and virtues of time averaging. *Paleobiology* 24, 287–304.

Kowalewski, M., Avila, S. G. E., Flessa, K. W., and Goodfriend, G. A. 2000. Dead delta's former productivity: two trillion shells at the mouth of the Colorado River. *Geology* 12, 1059–1062.

LaJeunesse, T., Reyes-Bonilla, H., and Warner, M. E. 2007. Spring bleaching among *Pocillopora* in the Sea of Cortez, eastern Pacific. *Coral Reefs* 26, 265–270.

Lancaster, N. 1995. *Geomorphology of Desert Dunes*. New York, Routledge. 290 p.

Lancaster, N., and Tchakerian, V. P. 1996. Geomorphology and sediments of sand ramps in the Mojave Desert. *Geomorphology* 17, 151–165.

Lancaster, N., and Tchakerian, V. P. 2003. *Late Quaternary Eolian Dynamics, Mojave Desert, California*. Geological Society of America, Special Paper 368, pp. 231–249.

Lara-Lara, J. R., Millán-Núnez, R., Lara-Osorio, J. L., and Bazán-Guzmán, C. 1993. Phytoplankton productivity and biomass by size classes in the Central Gulf of California during Spring 1985. *Ciencias Marinas* 19, 137–154.

Lavín, M. F., and Sanchez, S. 1999. On how the Colorado River affected the hydrography of the upper Gulf of California. *Continental Shelf Research* 19, 1545–1560.

Lavín, M. F., Gaxiola-Castro, G., Robles, J. M., and Richter, K. 1995. Winter water masses and nutrients in the northern Gulf of California. *Journal of Geophysical Research* 100, 8587–8605.

Lechuga-Deveze, C. H., Morquecho-Escamilla, M. L., Reyes-Salinas, A., and Hernandez-Alfonso, J. R. 2000. Environmental natural disturbances in Bahía Concepción, Gulf of California. In *Aquatic Ecosystems of Mexico: Status and Scope*. Edited by S.G.L.M. Munawar, I. F. Munawar, and D. F. Malley. Leiden, The Netherlands, Backhuys Publishers, pp. 245–255.

Ledesma-Vázquez, J. 2002. A gap in the Pliocene invasion of seawater, Gulf of California. *Revista Mexicana de Ciencias Geológicas* 19, 145–151.

Ledesma-Vázquez, J., and Johnson, M. E. 2001. Miocene-Pleistocene tectono-sedimentary evolution of Bahía Concepción region, Baja California Sur (México). *Sedimentary Geology* 144, 83–96.

Ledesma-Vázquez, J., Berry, R. W., Johnson, M. E., and Gutiérrez-Sanchez, S. 1997. El Mono chert: a shallow-water chert from the Pliocene Infierno Formation, Baja California Sur, Mexico. In *Pliocene Carbonates and Related Facies Flanking the Gulf of California, Baja California, Mexico*. Edited by M. E. Johnson and J. Ledesma-Vázquez. Geological Society of America Special Paper 318, pp. 73–81.

Ledesma-Vázquez, J., Carreño, A. L., Staines-Urias, F., and Johnson, M. E. 2006. The San Nicolás Formation: a proto-gulf extensional-related new lithostratigraphic unit at Bahía San Nicolás, Baja California Sur, Mexico. *Journal of Coastal Research* 22, 801–811.

Ledesma-Vázquez, J., Johnson, M. E., Backus, D. H., and Mirabal-Davila, C. 2007a. Coastal evolution from transgressive barrier deposit to marine terrace on Isla Coronados, Baja California Sur, México. *Ciencias Marinas* 33, 335–351.

Ledesma-Vázquez, J., Montiel-Boehringer, A. Y., Backus, D. H., Johnson, M. E., and Ferández-Díaz, V. Z. 2007b. Armored mud balls in tidal environments, Pliocene in the Gulf of California. In *Abstracts for the Fourth European Meeting on the Palaeontology and Stratigraphy of Latin America*. Edited by E.

Díaz-Martínez and I. Rábano. Cuademos del Museo Geomine-ro 8. Madrid, Instituto Geológico y Minero de España, pp. 235–238.

Lepley, L. K., Vonder-Haar, S. P., Hendrickson, J. R., and Calderón-Riverol, G. 1975. Circulation in the northern Gulf of California from orbital photographs and ship investigations. *Ciencias Marinas* 2, 86–93.

Leyte-Morales, G. E. 1995. Primer registro de *Gardineroseris planulata* (Dana, 1846) (Anthozoa: Scleractinia) en México. *XIII Congreso Nacional de Zoología*, Michoacan, México, pp. 162–163.

Libbey, L. K., and Johnson, M. E. 1997. Upper Pleistocene rocky shores and intertidal biotas on the Gulf of California at Playa La Palmita (Baja California Sur, Mexico). *Journal of Coastal Research* 13, 216–225.

Lirman, D., Glynn, P. W., Baker, A. C., and Leyte-Morales, G. E. 2001. Combined effects of three sequential storms on the Huatulco coral reef tract, Mexico. *Bulletin of Marine Science* 69, 267–278.

Lluch-Cota, S. E., and 29 coauthors. 2006. The Gulf of California: review of ecosystem status and sustainability challenges. *Progress in Oceanography* 73, 1–26.

Lonsdale, P., and Becker, K. 1985. Hydrothermal plumes, hot springs, and conductive heat flow in the Southern Trough of Guaymas Basin. *Earth and Planetary Science Letters* 73, 211–225.

López-Espinosa de los Monteros, R. 2002. Evaluating ecotourism in natural protected areas of La Paz Bay, Baja California Sur, México: Ecotourism or nature-based tourism? *Biodiversity and Conservation* 11, 1539–1550.

López-Hernandez, A., Garcia-Estrada, G., and Arellano-Guadarrama, F. 1994. Geological and geophysical studies at Las Tres Virgenes, B.C.S., Mexico. *Transactions of the Geothermal Resources Council* 18, 275–280.

López-Mendilaharsu, S., Gardner, C., and Riosmena-Rodríguez, R. 2005. Identifying critical foraging habitats of the green turtle (*Chelonia mydas*) along the Pacific Coast of the Baja California peninsula, México. *Aquatic Conservation: Marine and Freshwater Ecosystems* 15, 259–269.

López-Pérez, R. A. 2005. The Cenozoic hermatypic corals in the eastern Pacific: history of research. *Earth-Science Reviews* 75, 67–87.

López-Pérez, R. A., and Hernández-Ballesteros, L. M. 2004. Coral community structure and dynamics in the Huatulco area, western México. *Bulletin of Marine Science* 75, 453–472.

López-Pérez, R. A., Mora-Perez, M. G., and Leyte-Morales, G. E. 2007. Coral recruitment at Bahías de Huatulco, western México: implications for coral community structure and dynamics. *Pacific Science* 61, 355–369.

Lyle, M. W., and Ness, G. E. 1991. The opening of the southern Gulf of California. In *The Gulf and Peninsular Province of the Californias*. Edited by J. P. Dauphin and B. R. T. Simoneit. American Association of Petroleum Geologists Memoir 47, pp. 403–423.

Macias-Carranza, V. A. 1999. Pigmentos fitoplanotonicos por HPLC en el Golfo de California: Periodo Verano-Invierno 1996–1997. M.S. thesis, Universidad Autónoma de Baja California. 163 pp.

Marinone, S. G., and Ripa, P. 1988. Geostrophic flow in the Guaymas Basin, central Gulf of California. *Continental Shelf Research* 8, 159–166.

Marrack, E. 1999. The relationship between water motion and living rhodolith beds in the southwestern Gulf of California, Mexico. *Palaios* 14, 159–171.

Martin, S., Clavier, J., Chauvaud, L., and Thouzeau, G. 2007. Community metabolism in temperate maerl beds. II. Nutrient fluxes. *Marine Ecology Progress Series* 335, 31–41.

Martínez-Gutiérrez, G., and Sethi, P. S. 1997. Miocene-Pleistocene sediments within the San José del Cabo Basin. In *Pliocene Carbonates and Related Facies Flanking the Gulf of California, Baja California, México*. Edited by M. E. Johnson and J. Ledesma-Vázquez. Geological Society of America Special Paper 318, pp. 141–166.

Mateo-Cid, L. E., Sanchez-Rodríguez, I., Rodríguez-Montesinos, Y. E., and Casas-Valdez, M. M. 1993. Study on benthic marine algae of Bahía Concepción, B.C.S., Mexico. *Ciencias Marinas* 19, 41–60.

Mayer, L., and Vincent, K. R. 1999. Active tectonics of the Loreto area, Baja California Sur, Mexico. *Geomorphology* 27, 243–255.

McCune, B., and Grace, J. B. 2002. *Analysis of Ecological Communities*. Gleneden Beach, Ore., MjM Software. 210 p.

McDougall, K., Poore, R. Z., and Matti, J. C. 1999. Age and paleoenvironment of the Imperial Formation near San Gorgonio Pass, southern California. *Journal of Foraminiferal Research* 29, 4–25.

McFall, C. C. 1968. Reconnaissance geology of the Concepción Bay area, Baja California, Mexico. *Geological Sciences, Stanford University Publications* 10(5), 1–25.

McLean, H. 1989. Reconnaissance geology of a Pliocene marine embayment near Loreto, Baja California Sur, Mexico. In *Geologic Studies in Baja California*. Edited by P. L. Abbott. Pacific Section, Society of Economic Paleontologists and Mineralogists 63, pp. 17–25.

Medina-Lopez, M. A. 1999. Estuctura de la criptofauna asociada a mantos de rodolitos en el suroeste del Golfo de California, Mexico. B.S. thesis, Universidad Autonoma de Baja California Sur.

Medina-Rosas, P. 2006. Los corales hermatípicos de Mazatlán, Sinaloa, México. *Ciencia y Mar* 10, 13–17.

Merrifield, M. A., and Winant, C. D. 1989. Shelf circulation in the Gulf of California: a description of the variability. *Journal of Geophysical Research* 94(C12), 18133–18160.

Millán-Núñez, R., Cajal-Medrano, R., Santamaría-del-Angel, E., and Millán-Núñez, E. 1993. Primary productivity and chlorophyll *a* in the central part of the Gulf of California (autumn 1987). *Ciencias Marinas* 19, 29–40.

Millán-Núñez, R., Santamaría-del-Angel, E., Cajal-Medrano, R., and Barocio-León, O. C. 1999. The delta of the Colorado River: an ecosystem with high primary productivity. *Ciencias Marinas* 25, 509–524.

Miller, K. G., Kominz, M. A., Browning, J. V., Wright, J. D., Mountain, G. S., Katz, M. E., Sugarman, P. J., Cramer, B. S., Christie-Blick, N., and Pekar, S. 2005. The Phanerozoic record of sea-level change. *Science* 310, 1293–1298.

Miranda-Reyes, F., Reyes-Coca, S., and Garda-López, J. 1990. *Climatología de la region noroeste de México. Parte I. Precipitación*. Rep. Tec. EBA No. 3, CICESE, Ensenada, Baja California, México. 160 p.

Mitchell, A. J., and Collins, K. J. 2004. Understanding the distribution of maerl, a calcareous seaweed, off Dorset, U.K. *Proceedings of the Second International Symposium on GIS/Spatial*

Analyses in Fishery and Aquatic Sciences, University of Sussex, Brighton, U.K., pp. 65–82.

Möbius, K. 1883. The oyster and oyster-culture. In *United States Commission of Fish and Fisheries, Report of the Commissioner for 1880*. Part VIII. Washington, D.C., pp. 683–751.

Mora-Pérez, M. G. 2005. Biología reproductiva del coral *Porites panamensis* Verrill 1866 (Anthozoa: Scleractinia), en bahía de La Paz, Baja California Sur, México. M.Sc. thesis, Centro Interdisciplinario de Ciencias Marinas, La Paz.

Muhs, D. R., Kennedy, G. L., and Rockwell, T. K. 1994. Uranium-series ages of marine terrace corals from the Pacific coast of North America and implications for last-interglacial sea level history. *Quaternary Research* 42, 72–87.

Muller-Parker, G., and D´Elia, C. F. 1997. Interactions between corals and their symbiotic algae. In *Life and Death in Coral Reefs*. Edited by C. Birkeland. New York, Chapman and Hall, pp. 96–113.

Nagy, E. A., and Stock, J. N. 2000. Structural controls on the continent-ocean transition in the northern Gulf of California. *Journal of Geophysical Research* 105, 16251–16269.

Nebelsick, J. H., and Kroh, A. 2002. The stormy path from life to death assemblages: the formation and preservation of mass accumulations of fossil sand dollars. *Palaios* 17, 378–393.

Nelson, C. S. 1988. An introductory perspective on nontropical shelf carabonates. *Sedimentary Geology* 60, 3–12.

Nicholson, C., Sorlien, C. C., Atwater, T., Crowell, J. C., and Luyendyk, B. P. 1994. Microplate capture, rotation of the western Transverse Ranges, and initiation of the San Andreas transforma as a low-angle fault system. *Geology* 22, 491–495.

Ortlieb, L. 1991. Quaternary vertical movements along the coasts of Baja California and Sonora. In *The Gulf and Peninsular Provinces of the Californias*. Edited by J. P. Dauphin and B. R. T. Simoneit. American Association of Petroleum Geologists Memoir 47, pp. 447–480.

Oskin, M., and Stock, J. 2003. Marine incursion synchronous with plate boundary localization in the Gulf of California. *Geology* 31, 23–26.

Oskin, M., Stock, J., and Martin-Barajas, A. 2001. Rapid localization of Pacific–North America plate motion in the Gulf of California. *Geology* 29, 459–462.

Oxburgh, E. R., and O'Nions, R. K. 1987. Helium loss, tectonics, and the terrestrial heat budget. *Science* 237, 1583–1588.

Pacheco-Ruíz, I., and Zertuche-Gonzaléz, J. A. 2002. Red algae (Rhodophyta) from Bahía de Los Angeles, Gulf of California, Mexico. *Botanica Marina* 45, 465–470.

Paden, C. A., Abbott, M. R., and Winant, C. D. 1991. Tidal and atmospheric forcing of the upper ocean in the Gulf of California. 1. Sea surface temperature variability. *Journal Geophysical Research* 96, 18337–18359.

Palmer, R. H. 1928. Geology of southern Oaxaca, México. *Journal of Geology* 36, 718–734.

Pandolfi, J. M. 1996. Limited membership in Pleistocene reef coral assemblages from the Huon Peninsula, Papua New Guinea: constancy during global change. *Paleobiology* 22, 152–176.

Parés-Sierra, A., Mascarenhas, A., Marinone, S. G., and Castro, R. 2003. Temporal and spatial variation of the surface winds in the Gulf of California. *Geophysical Research Letters* 30(6), 1312–1316.

Paul-Chavez, L., and Riosmena-Rodríguez, R. 2000. Floristic and biogeographical trends in seaweed assemblages from a

subtropical insular island complex in the Gulf of California. *Pacific Science* 54, 137–147.

Pérez-España, H., Galván-Magaña, F., and Abitia-Cárdenas, L. A. 1996. Temporal and spatial variations in the structure of the rocky reef fish community of the southwest Gulf of California, Mexico. *Ciencias Marinas* 22, 273–294.

Piazza, M., and Robba, E. 1998. Autochthonous biofacies in the Pliocene Loreto Basin, Baja California Sur, Mexico. *Rivista Italiana di Paleontologia e Stratigrafia* 104, 227–262.

Pichler, T., Veizer, J., and Hall, G. E. M. 1999. Natural input of arsenic into a coral-reef ecosystem by hydrothermal fluids and its removal by Fe(III) oxyhydroxides. *Environmental Science and Technology* 33, 1373–1378.

Pichler, T., Amend, J., Garey, J., Hallock, P., Hsia, N., Karlen, D., McCloskey, B., Meyer-Dombard, D., and Price, R. 2006. A natural laboratory to study arsenic geobiocomplexity. *EOS Transactions of the American Geophysical Union* 87, 221–225.

Pitt, J., Luecke, D., Cohen, M., Glenn, E., and Valdes, C. 2000. Two nations, one river: managing ecosystems conservation in the Colorado River delta. *Natural Resources Journal* 40, 819–864.

Poreda, R. J., Jenden, P. D., Kaplan, I. R., and Craig, H. 1986. Mantle helium in Sacramento basin natural gas wells. *Geochimica et Cosmochimica Acta* 50, 2847–2853.

Portugal, E., Birkle, P., Barragan, R. M., Arellano, V. M., Tello, E., and Tello, M. 2000. Hydrochemical-isotopic and hydrogeological conceptual model of the Las Tres Vírgenes geothermal field, Baja California Sur, México. *Journal of Volcanology and Geothermal Research* 101, 223–244.

Price, R. E., and Pichler, T. 2005. Distribution, speciation and bioavailability of arsenic in a shallow-water submarine hydrothermal system, Tutum Bay, Ambitle Island, PNG. *Chemical Geology* 224, 122–135.

Pride, C. 1997. An evaluation and application of paleoceanographic proxies in the Gulf of California. Ph.D. diss., University of South Carolina. 196 p.

Prol-Ledesma, R. M. 2003. Similarities in the chemistry of shallow submarine hydrothermal vents. *Geothermics* 32, 639–644.

Prol-Ledesma, R. M., Canet, C., Melgarejo, J. C., Tolson, G., Rubio-Ramos, M. A., Cruz-Ocampo, J. C., Ortega-Osorio, A., Torres-Vera, M. A., and Reyes, A. 2002. Cinnabar deposition in submarine coastal hydrothermal vents, Pacific Margin of central Mexico. *Economic Geology* 97, 1331–1340.

Prol-Ledesma, R. M., Canet, C., Torres-Vera, M. A., Forrest, M. J., and Armienta, M. A. 2004. Vent fluid chemistry in Bahía Concepción coastal submarine hydrothermal system, Baja California, Mexico. *Journal of Volcanology and Geothermal Research* 137, 311–328.

Reed, C. 2006. Boiling points. *Nature* 439, 905–907.

Reyes-Bonilla, H. 1992. New records for hermatypic corals (Anthozoa: Scleractinia) in the Gulf of California, México, with an historical and biogeographical discussion. *Journal of Natural History* 26, 1163–1175.

Reyes-Bonilla, H. 1993a. Biogeografía y ecología de los corales hermatípicos (Anthozoa: Scleractinia) del Pacífico de México. In *Biodiversidad Marina y Costera de México*. Edited by S. Salazar-Vallejo and N. E. González. Chetumal, CONABIO/CIQRO, pp. 207–222.

Reyes-Bonilla, H. 1993b. The 1987 coral reef bleaching at Cabo Pulmo reef, Gulf of California, México. *Bulletin of Marine Science* 52, 832–837.

Reyes-Bonilla, H. 1998. *Pocillopora damicornis* (Anthozoa: Scler-actinia) in the Gulf of California, México. *Revista de Biologia Tropical* 46, 463–468.

Reyes-Bonilla, H. 2001. Effects of the 1997–1998 El Nino-Southern Oscillation on coral communities of the Gulf of California, México. *Bulletin of Marine Science* 69, 251–266.

Reyes-Bonilla, H. 2002. Checklist of valid names and synonyms of stony corals (Anthozoa: Scleractinia) from the Eastern Pacific. *Journal of Natural History* 36, 1–13.

Reyes-Bonilla, H. 2003. Coral reefs of the Pacific coast of México. In *Latin American Coral Reefs*. Edited by J. Cortés. Amsterdam: Elsevier, pp. 331–349.

Reyes-Bonilla, H., and Calderón-Aguilera, L. E. 1994. Parámetros poblacionales de *Porites panamensis* (Anthozoa: Scleractinia), en el arrecife Cabo Pulmo, México. *Revista de Biologia Tropical* 4, 121–128.

Reyes-Bonilla, H., and Calderón-Aguilera, L. E. 1999. Population density, distribution and consumption rates of three coralli-vores at Cabo Pulmo reef, Gulf of California, Mexico. *Marine Ecology* 20, 347–357.

Reyes-Bonilla, H., and Cruz-Piñón, G. 2000. Biogeography of the ahermatypic corals (Scleractinia) of the Mexican Pacific. *Ciencias Marinas* 26, 511–531.

Reyes-Bonilla, H., and López-Pérez, R. A. 1998. Biogeography of the stony corals (Scleractinia) of the Mexican Pacific. *Ciencias Marinas* 24, 211–224.

Reyes-Bonilla, H., Riosmena-Rodríguez, R., and Foster, M. S. 1997. Hermatypic corals associated with rhodolith beds in the Gulf of California. *Pacific Science* 51, 328–337.

Reyes-Bonilla, H., Pérez-Vivar, T. L., and Ketchum-Mejía, J. T. 1999. Distribución geográfica y depredación de *Porites lobata* (Anthozoa: Scleractinia) en la costa occidental de México. *Revista de Biología Tropical* 47, 273–279.

Reyes-Bonilla, H., Carriquiry, J. D., Leyte-Morales, G. E., and Cupul-Magaña, A. L. 2002. Effects of the El Niño-Southern Oscilla-tion and the anti-El Niño event (1997–1999) on coral reefs of the western coast of Mexico. *Coral Reefs* 21, 368–372.

Reyes-Bonilla, H., Calderón-Aquilera, L. E., Cruz-Piñon, G., Medina-Rosas, P., López-Pérez, R. A., Herrero-Pérezrul, M. D., Leyte-Morales, G. E., Cupul-Magaña, A. L., and Carriquiry-Beltrán, J. D. 2005a. *Atlas de los Corales Pétreos (Anthozoa: Scleractinia) del Pacific Mexicano*. Centro de Investigación Científica y de Educación Superior de Ensenada, Comisión Nacional para el Conocimiento y Uso de la Biodiversidad, Consejo Nacional de Ciencia y Tecnologia, Universidad de Gudalajara/Centro Uni-versitario de la Costa, Universidad del Mar. 1128 p.

Reyes-Bonilla, H., González-Azcárraga, A., and Rojas-Sierra, A. 2005b. Estructura de las asociaciones de las estrellas de mar (Asteroidea) en arrecifes rocosos del Golfo de California, México. *Revista de Biología Tropical* 53, 233–244.

Reyes-Bonilla, H., González-Romero, S., Cruz-Piñón, G., and Calderón-Aguilera, L. E. 2008. Corales pétreos. In *Bahía de Los Angeles: recursos naturales y comunidad. Línea base 2007*. Ed-ited by G. D. Danemann and E. Ezcurra. México City, Instituto Nacional de Ecología, Pronatura Noroeste, San Diego Natural History Museum, pp. 291–317.

Richmond, R. H. 1985. Variations in the population biology of *Po-cillopora damicornis* across the Pacific. *Proceedings of the Fifth International Coral Reef Symposium* 6, 101–106.

Richmond, R. H. 1987. Energetic relationships and biogeographical

differences among fecundity, growth and reproduction in the reef coral *Pocillopora damicornis*. *Bulletin of Marine Science* 4, 594–604.

Richmond, R. H. 1997. Reproduction and recruitment in coral: critical links in the persistence of reef. In *Life and Death of Cor-al Reefs*. Edited by C. E. Birkeland. New York, Chapman and Hall, pp. 175–196.

Ricketts, E. F., Calvin, J., and Hedgpeth, J. W. 1953. *Between Pacific Tides*. Stanford, Calif.: Stanford University Press. 502 p.

Riegl, B., and Piller, W. E. 2003. Possible refugia for reefs in times of environmental stress. *International Journal of Earth Science* 92, 520–531.

Riegl, B. M., Halfar, J., Purkis, S. J., and Godínez-Orta, L. 2007. Sedimentary facies of the eastern Pacific´s northernmost reef-like setting (Cabo Pulmo, Mexico). *Marine Geology* 236, 61–77.

Riosmena-Rodríguez, R. 2002. Taxonomy of the order Corallinales (Rhodophyta) in the Gulf of California, México. Ph.D. diss., La Trobe University. 200 p., 289 plates.

Riosmena-Rodríguez, R., Woelkerling, W. J., and Foster, M. S. 1999. Taxonomic reassessment of rhodolith-forming species of *Lithophyllum* (Corallinales, Rhodophyta) in the Gulf of Cali-fornia, Mexico. *Phycologia* 38, 401–417.

Riosmena-Rodríguez, R., Steller, D. L., Hinojosa-Arango, G., and Foster, M. S. In press. Reefs that rock and roll: biology and con-servation of rhodolith beds in the Gulf of California. In *Marine Biodiversity and Conservation in the Gulf of California*. Edited by R. Brusca. Tucson, University of Arizona Press and Sonoran Desert Museum.

Rivera, M. G., Riosmena-Rodríguez, R., and Foster, M. S. 2004. Age and growth of *Lithothamnion muelleri* (Corallinales, Rho-dophta) in the southwestern Gulf of California, Mexico. *Cien-cias Marinas* 30, 235–249.

Roberts, J. M., Wheeler, A. J., and Freiwald, A. 2006. Reefs of the deep: the biology and geology of cold-water coral ecosystems. *Science* 312, 543–547.

Roberts, N. C. 1989. *Baja California Plant Field Guide*. La Jolla, Natural History Publishing Co. 309 p.

Roberts, S., and Hirschfield, M. 2004. Deep-sea corals: out of sight but no longer out of mind. *Frontiers in Ecology and the Envi-ronment* 2, 123–130.

Roden, G. I. 1964. Oceanographic aspects of the Gulf of California. In *Marine Geology of the Gulf of California: A Symposium*. Edit ed by T. J. H. Van Andel and G. G. Shor. American Association of Petroleum Geologists Memoir 3, 30–58.

Roden, G. I., and Groves, G. W. 1959. Recent oceanographic obser-vations in the Gulf of California. *Journal of Marine Resources* 18, 10–35.

Rodger, K. A., ed. 2006. *Breaking Through: Essays, Journals, and Travelogues of Edward F. Ricketts*. Berkeley: University of Cali-fornia Press. 348 p.

Rodríguez, C. A., Flessa, K. W., and Dettman, D. L. 2001a. Effects of upstream diversion of Colorado River water on the estuarine bivalve mollusk *Mulinia coloradoensis*. *Conservation Biology* 15, 249–258.

Rodríguez, C. A., Flessa, K. W., Téllez-Duarte, M. A., Dettman, D. L., and Avila-Serrano, G. E. 2001b. Macrofaunal and isotropic estimates of the former extent of the Colorado River estuary, upper Gulf of California, México. *Journal of Arid Environments* 49, 183–193.

Rodríguez-Troncoso, A. P. 2006. Ciclo reproductivo de tres

especies de corales formadoras de arrecife en Bahía La Entrega, Oaxaca, México. M.Sc. thesis, Universidad Autónoma de Baja California, Ensenada.

Rowell, K. 2006. Isotopic logs of the Sea of Cortez: oxygen and carbon stable isotopes in otoliths from marine fish record the impact of diverting the Colorado River from the sea. Ph.D. diss., University of Arizona, Tucson.

Rowell, K., Flessa, K. W., Dettman, D. L., and Roman, M. 2005. The importance of Colorado River flow to nursery habitats of the gulf curvina (*Cynoscion othonopterus*). *Canadian Fisheries and Aquatic Sciences* 72, 2874–2885.

Russell, P., and Johnson, M. E. 2000. Influence of seasonal winds on coastal carbonate dunes from the Recent and Plio-Pleistocene at Punta Chivato (Baja California Sur, Mexico). *Journal of Coastal Research* 16, 709–723.

Sáenz-Arroyo, A., Roberts, C. M., Torre, J., Cariño-Olvera, M., and Enríquez-Andrade, R. R. 2005. Rapidly shifting environmental baselines among fishers of the Gulf of California. *Proceedings of the Royal Society, series B* 272 1957–1962.

Sáenz-Arroyo, A., Roberts, C. M., Torre, J., Cariño-Olvera, M., and Hawkins, J. P. 2006. The value of evidence about past abundance: marine fauna of the Gulf of California through the eyes of 16th to 19th century travellers. *Fish and Fisheries* 7, 128–146.

Sala, E., Aburto-Oropeza, O., Paredes, G., Parra, I., Barrera, J. C., and Dayton, P. K. 2002. A general model for designing networks of marine reserves. *Science* 298, 1991–1993.

Sala, E., Aburto-Oropeza, O., Reza, M., Paredes, G., and López-Lemus, L. G. 2004. Fishing down coastal food webs in the Gulf of California. *Fisheries* 29, 19–25.

Salinas-Prieto, J. C., Montiel-Escobar, J., Sánchez-Rojas, E., Díaz-Salgado, C., De la Calleja, A., Barajas-Nigoche, D., Dorantes-Salgado, E., Jiménez-Gonzaga, A., and Amescua-Torres, N. 2007. Carta Geológica de México, escala 1:2,000,000. Sexta edición: Servicio Geológio Mexicano.

Santamaría-del-Angel, E., Alvarez-Borrego, S., and Müller-Karger, F. E. 1994a. Gulf of California biogeographic regions based on coastal zone color scanner imagery. *Journal of Geophysical Research* 99(C4), 7411–7421.

Santamaría-del-Angel, E., Alvarez-Borrego, S., and Müller-Karger, F. E. 1994b. The 1982–1984 El Niño in the Gulf of California as seen in coastal zone color scanner imagery. *Journal of Geophyscial Research* 99(C4), 7423–7431.

Santamaría-del-Angel, E., Alvarez-Borrego, S., Millán-Núñez, R., and Müller-Karger, F. E. 1999. On the weak effect of summer upwelling on the phytoplankton biomass of the Gulf of California. *Revista de la Sociedad Mexicana de Historia Natural* 49, 207–212.

Sawlan, M. G. 1991. Magmatic evolution of the Gulf of California rift. In *The Gulf and Peninsular Province of the Californias*. Edited by J. P. Dauphin and B. R. T. Simoneit. American Association of Petroleum Geologists Memoir 47, pp. 217–229.

Sawlan, M. G., and Smith, J. G. 1984. Petrologic characteristics, age and tectonic setting of Neogene volcanic rocks in northern Baja California Sur, Mexico. In *Geology of the Baja California Peninsula*. Vol. 39. Edited by V. A. Frizzell Jr. Los Angeles, Pacific Section, Society of Economic Paleontologists and Mineralogists, pp. 219–236.

Schopf, J. W., Kudryavtsev, A. B., Agresti, D. G., Wdowiak, T. J., and Czaja, A. D. 2002. Laser-Raman imagery of Earth's earliest fossils. *Nature* 416, 73–76.

Sedlock, R. L., Ortega-Gutiérrez, F., and Speed, R. C. 1993. *Tectonostratigraphic terranes and tectonic evolution of Mexico*. Geological Society of America Special Paper 278. 153 p.

SEMARNAT. 2001. Protección ambiental: Especies nativas de México de flora y fauna silvestres: Categorías de riesgo y especificaciones para su inclusión, exclusión o cambio: Lista de especies en riesgo. *Diario Oficial de la Federación*, Marzo 6 2002, pp. 1–85.

SEMARNAT. 2007. *Plan de Ordenamiento Ecologico Marino del Golfo de California*. SEMARNAT, Mexico.

Severinghaus, J., and Atwater, T. 1990. Cenozoic geometry and thermal state of the subducting slabs beneath western North America. In *Basin and Range Extensional Tectonics Near the Latitude of Las Vegas, Nevada*. Edited by B. Wernicke. Geological Society of America Memoir 176, pp. 1–22.

Sewell, A., Johnson, M. E., Backus, D., and Ledesma-Vázquez, J. 2007. Rhodolith detritus impounded by a coastal dune on Isla Coronados, Gulf of California. *Ciencias Marinas* 33, 483–494.

Shafiquallah, M., Damon, P. E., Lynch, D. J., Reynolds, S. J., Rehrig, W. A., and Raymonds, R. H. 1980. K/Ar geochronology and geologic history of the southwestern Arizona and adjacent areas. *Arizona Geological Society Digest* 121, 201–260.

Shaw, C. A. 1981. The Middle Pleistocene El Golfo local fauna from northwestern Sonora, Mexico. M.S. thesis, California State University, Long Beach. 141 p.

Shepard, F. P. 1950. Submarine topography of the Gulf of California. In *1940 E. W. Scripps Cruise to the Gulf of California*. Part 3. By C. A. Anderson et al. Geological Society of America Memoir 43. 32 p.

Siever, R. 1983. Evolution of chert at active and passive continental margins. In *Siliceous Deposits in the Pacific Region*. Edited by A. Iijima, J. R. Hein, and R. Siever. Developments in Sedimentology 36, pp. 7–25.

Simian, M. E., and Johnson, M. E. 1997. Development and foundering of the Pliocene Santa Ines Archipelago in the Gulf of California: Baja California Sur, Mexico. In *Pliocene Carbonates and Related Facies Flanking the Gulf of California, Baja California, Mexico*. Edited by M. E. Johnson and J. Ledesma-Vázquez. Geological Society of America Special Papers 318, pp. 25–38.

Simpson, J. H., Souza, A. J., and Lavín, M. F. 1994. Tidal mixing in the Gulf of California. In *Mixing and Transport in the Environment*. Edited by K. Beven, P. Chatwin and J. Millbank. Cathy Allen Memorial Volume, John Wiley, Lancaster University, pp. 169–179.

Sirkin, L., Szabo, B. J., Padilla, A. G., Pedrin, A. S., and Diaz, R. E. 1990. Uranium-series ages of marine terraces, La Paz peninsula, Baja California Sur, México. *Coral Reefs* 9, 25–30.

Skudder, P. A., Backus, D. H., Goodwin, D. H., and Johnson, M. E. 2006. Sequestration of carbonate shell material in coastal dunes on the Gulf of California (Baja California Sur, Mexico). *Journal of Coastal Research* 22, 611–624.

Smith, J. T. 1984. Miocene and Pliocene marine mollusks and preliminary correlations, Vizcaíno Peninsula to Arroyo la Purísima, northwestern Baja California Sur, Mexico. In *Geology of the Baja California Peninsula*. Vol. 39. Edited by V. A. Frizzell Jr. Los Angeles, Pacific Section, Society of Economic Paleontologists and Mineralogists, pp. 197–217.

Smith, J. T. 1991a. *Cenozoic giant Pectinids from California and the Tertiary Caribbean Province: Lyropecten, "Macrochlamis," Vertipecten, and Nodipecten species*. U.S. Geological Survey Professional Paper 1391. 155 p., 38 plates

Smith, J. T. 1991b. Cenozoic marine mollusks and paleogeography of the Gulf of California. In *The Gulf and Peninsular Provices of the Californias*. Edited by J. P. Dauphin and B. R. T. Simoneit. American Association of Petroleum Geologists Memoir 47, pp. 637–666.

Solis-Marín, F. A., Reyes-Bonilla, H., Herrero-Pérezrul, M. D., Arizpe-Cobarrubias, O., and Laguarda-Figueras, A. 1997. Systematics and distribution of the echinoderms from Bahía de la Paz. *Ciencias Marinas* 23, 249–263.

Southward, A. J., Southward, E. C., Dando, P. R., Hughes, J. A., Kennicutt II, M. C., Alcala-Herrera, J., and Leahy, Y. 1997. Behaviour and feeding of the Nassariid gastropod *Cyclope Neritea*, abundant at hydrothermal brine seeps off Milos (Aegean Sea). *Journal of the Marine Biological Association of the United Kingdom* 77, 753–771.

Squires, D. F. 1959. Results of the Puritan-American Museum of Natural History expedition to western México. 7. Corals and coral reefs in the Gulf of California. *Bulletin of the American Museum of Natural History* 118, 370–431.

Stanley, S. M. 1968. Post-Paleozoic adaptive radiation of infaunal bivalve molluscs: a consequence of mantle fusion and siphon formation. *Journal of Paleontology* 42, 214–229.

Steinbeck, J., and Ricketts, E. F. 1941. *Sea of Cortez: A Leisurely Journal of Travel and Research*. New York, Viking Press. 598 p.

Steller, D. L. 2003. Rhodoliths in the Gulf of California: growth, demography, disturbance and effects on population dynamics of catarina scallops. Ph.D. diss., University of California, Santa Cruz. 143 p.

Steller, D. L., and Foster, M. S. 1995. Environmental factors influencing distribution and morphology of rhodoliths in Bahía Concepción, B.C.S., Mexico. *Journal of Experimental Marine Biology and Ecology* 194, 201–212.

Steller, D. L., Riosmena-Rodríguez, R., Foster, M. S., and Roberts, C. A. 2003. Rhodolith bed diversity in the Gulf of California: the importance of rhodolith structure and consequences of disturbance. *Aquatic Conservation-Marine and Freshwater Ecosystems* 13, S5–S20.

Steller, D. L., Hernandez, M., Riosmena-Rodríguez, R., and Cabello-Pasini, A. 2007a. Effect of temperature on photosynthesis, growth and calcification rates of the free-living coralline alga *Lithophyllum margaritae*. *Ciencias Marinas* 33, 441–456.

Steller, D. L., Foster, M. S., Riosmena-Rodríguez, R. 2007b. Sampling and monitoring rhodolith beds. In *Sampling Biodiversity in Coastal Communities: NaGISA Protocols for Seagrass and Macroalgal Habitats*. Edited by K.I.P.R. Rigy and Y. Shirayama. Kyoto, Kyoto University Press, pp. 93–97.

Stetter, K. O. 2006. Hyperthermophiles in the history of life. *Philosophical Transactions of the Royal Society B: Biological Sciences* 361, 1837–1843.

Stock, J. M., and Hodges, K. V. 1989. Pre-Pliocene extension around the Gulf of California and the transfer of Baja California to the Pacific plate. *Tectonics* 8, 99–115.

Stock, J. M., Martin-Barajas, A., and Tellez-Duarte, M. 1996. Early rift sedimentation and structure along the NE margin of Baja California. *American Association of Petroleum Geologists Field Conference Guide, Pacific Section SEPM* 80, 337–372.

Suárez-Bosche, N., Suárez-Bosche, K., and Suárez A., M. C. 2000. Submarine geothermal systems in Mexico. *Proceedings of the World Geothermal Congress 2000*, pp. 3889–3893.

Sykes, G. G. 1937. *The Colorado Delta*. Carnegie Institution Publication 460. 193 p.

Tarasov, V. G., Propp, M. V., Propp, L. N., Zhirmunsky, A. V., Namsaraev, B. B., Gorlenko, V. M., and Starynin, D. A. 1990. Shallow-water gasohydrothermal vents of Ushishir Volcano and the ecosystem of Kraternaya Bight (The Kurile Islands). *Marine Ecology* 11, 1–23.

Tarasov, V. G., Gebruk, A. V., Shulkin, V. M., Kamenev, G. M., Fadeev, V. I., Kosmynin, V. N., Malakhov, V. V., Starynin, D. A., and Obzhirov, A. I. 1999. Effect of shallow-water hydrothermal venting on the biota of Matupi Harbour. *Continental Shelf Research* 19, 79–116.

Tarasov, V. G., Gebruk, A. V., Mironov, A. N., and Moskalev, L. I. 2005. Deep-sea and shallow-water hydrothermal vent communities: Two different phenomena? *Chemical Geology* 224, 5–39.

Tchakerian, V. P. 1991. Late Quaternary Aeolian geomorphology of the Dale Lake Sand Sheet, southern Mojave Desert, California. *Physical Geography* 12(4), 347–369.

Tchakerian, V. P., and Lancaster, N. 2002. Late Quaternary arid/humid cycles in the Mojave Desert and western Great Basin of North America. *Quaternary Science Reviews* 21, 799–810.

Téllez-Duarte, M. A., Avila, G. S., and Flessa, K. W. 2001a. Los concheros arqueológicos del Delta del Colorado y el uso sustentable de los recursos naturals. *Primera Reunión Binacional: Balances y Perspectivas de la Baja California Prehispánica e Hispánica Memoria* (editada en disco compacto). Mexicali, Baja California, INAH.

Téllez-Duarte, M. A., Avila-Serrano, G. E., and Flessa, K. W. 2001b. Some observations on shell middens from the Colorado River delta area. *34th Annual Meeting Abstracts, Western Society of Malacology*, p. 34.

Téllez-Duarte, M. A., Avila, G. E., and Flessa, K. W. 2005. Archaeological shell middens in the Colorado delta: an option for the use of the Biosphere Reserve of the Upper Gulf of California. *Pacific Coast Archaeological Society Quarterly* 37(4), 80–86.

Tershy, B. R., Bourillón, L., Metzler, L., and Barnes, J. 1999. A survey of ecotourism on islands in northwestern México. *Environmental Conservation* 26, 212–217.

Thomas, D. S. G. 1997. Sand seas and Aeolian bedforms. In *Arid Zone Geomorphology: Process, Form and Change in Drylands*. 2nd ed. Edited by D. S. G. Thomas. Chichester, U.K., John Wiley and Sons, pp. 373–412.

Thomson, D. A., Findley, L. T., and Kerstitch, A. N. 2000. *Reef Fishes of the Sea of Cortez*. Rev. ed. Austin, University of Texas Press.

Thompson, R. W. 1968. *Tidal Flat Sedimentation on the Colorado River Delta, Northwestern Gulf of California*. Geological Society of America Memoir 107. 133 p.

Torres-Orozco, E. 1993. Análisis volumétrico de las masas de agua del Golfo de California. M.Sc. thesis, CICESE, Ensenada Baja California, México. 80 p.

Trager, G. C., and De Niro, M. J. 1990. Chemoautotrophic sulfur bacteria as a food source for mollusks at intertidal hydrothermal vents: evidence from stable isotopes. *Veliger* 33(4), 359–362.

Ulloa, R., Torre, J., Bourillon, L., Gondor, A., and Alcantar, N. 2006. *Planeación Ecoregional para la Conservación Marina: Gulfo de California y Costa Occidental de Baja California Sur*. México, D.F., The Nature Conservancy. 153 p.

Umhoefer, P. J., Dorsey, R. J., and Renne, P. 1994. Tectonics of the Pliocene Loreto basin, Baja California Sur, Mexico, and evolution of the Gulf of California. *Geology* 22, 649–652.

Umhoefer, P. J., Schwennicke, T., Del Margo, M. T., Ruiz-Geraldo, G., Ingle, J. C., and McIntosh, W. 2007. Transtensional fault-termination basins: an important basin type illustrated by the Pliocene San José Island basin and related basins in the southern Gulf of California, Mexico. *Basin Research* 19, 297–322.

Underwood, A. J. 1997. *Experiments in Ecology: Their Logical Design and Interpretation Using Analysis of Variance*. Cambridge, Cambridge University Press. 50 p.

Valdez-Holguín, J. E. 1986. Distribución de la biomasa y productividad del fitoplancton en el Golfo de California durante el evento de El Niño 1982–1983. M.Sc. thesis, CICESE, Ensenada Baja California, México. 90 p.

Vaughan, T. W. 1917. The coral reef fauna of Carrizo Creek, Imperial County, California, and its significance. *United States Geological Survey Professional Paper* 98, 355–386.

Vázquez-Elizondo, R. M. 2005. Revalución taxonomic de *Lithophyllum bracchiatum* (Heydrich) ME Lemoine (Rhodophyta: Corallinales) para el suroeste del Golfo de California. Tesis de Licenciatura, UABCS, La Paz, México. 52 p.

Veron, J. E. N. 1995. *Corals in Time and Space*. Ithaca, N.Y.: Comstock and Cornell University Press. 321 p.

Vidal, V. M. V., Vidal, F. V., Isaacs, J. D., and Young, D. R. 1978. Coastal submarine hydrothermal activity off Northern Baja California. *Journal of Geophysical Research* 83(B4), 1757–1774.

Viesca-Lobatón, C., Balart, E. F., González-Cabello, A., Mascareñas-Osorio, I., Aburto-Oropeza, O., Reyes-Bonilla, H., and Torreblanca, E. 2008. Peces arrecifales. In *Bahía de Los Angeles: recursos naturales y comunidad. Línea base 2007*. Edited by G. D. Danemann and E. Ezcurra. México City, Instituto Nacional de Ecología, Pronatura Noroeste, San Diego Natural History Museum, pp. 385–427.

Villalard-Bohnsack, M., and Harlin, M. M. 1992. Seasonal distribution and reproductive status of macroalgae in Narragansett Bay and associated waters, Rhode Island, U.S.A. *Botanica Marina* 35, 205–214.

Villarreal-Chávez, G., Garcia-Dominguez, F., Corea, F. Y., and Castro-Castro, N. 1999. Note on the geographic distribution of *Chinoe cortezi* (Carpenter, 1864) (Mollusca: Pelecypoda: Veneridae). *Ciencias Marinas* 25, 153–159.

Villarreal-Cavazos, A., Reyes-Bonilla, H., Bermúdez-Almada, B., and Arizpe-Covarrubias, O. 2000. Los peces del arrecife de Cabo Pulmo, Golfo de California, México: Lista sistemática y aspectos de abundancia y biogeografía. *Revista de Biología Tropical* 48, 413–424.

Vizcaíno-Ochoa, V. E. 2003. Biología reproductiva de tres especies de corales formadores de arrecifes en Bahía Banderas, México. M.Sc. thesis, Universidad Autónoma de Baja California, Ensenada.

Wadge, G., and Quarmby, N. 1988. Geological remote sensing of rocky coasts. *Geological Magazine* 125, 495–505.

Walker, T. R., and Thompson, R. W. 1968. Late Quaternary geology of the San Felipe Area, Baja California, Mexico. *Journal of Geology* 76, 479–485.

Wang, C., and Fiedler, P. C. 2006. ENSO variability and the eastern tropical Pacific: a review. *Progress in Oceanography* 69, 239–266.

Waters, M. R. 1983. Late Holocene lacustrine chronology and archeology of ancient Lake Cahuilla, California. *Quaternary Research* 19, 373–387.

Watkins, R. 1990. Paleoecology of a Pliocene rocky shoreline, Salton Trough region, California. *Palaios* 5, 167–175.

Welhan, J. A. 1988. Origins of methane in hydrothermal systems. *Chemical Geology* 71, 183–198.

Wellington, G. M. 1982. Depth zonation of corals in the Gulf of Panamá: control and facilitation by resident reef fishes. *Ecological Monographs* 52, 223–241.

Wiggins, I. L. 1980. *Flora of Baja California*. Stanford, Calif., Stanford University Press. 1025 p.

Wilson, I., and Rocha, V. 1955. Geology and mineral deposit of the Boleo Copper District Baja California, México. *Geological Survey Professional Paper* 273, 37–39.

Wilson, S., Blake, C., Berges, J. A., and Maggs, C. A. 2004. Environmental tolerances of free-living coralline algae (maerl): implications for European marine conservation. *Biological Conservation* 120, 279–289.

Winker, C. D., and Kidwell, S. M. 1996. Stratigraphy of a marine rift basin: Neogene of the western Salton Trough, California. *American Association of Petroleum Geologists Field Conference Guide, Pacific Section SEPM* 80, 295–336.

Woelkerling, W. J., and Lamy, D. 1998. *Non-geniculate Coralline Red Algae and the Paris Muséum: Systematics and Scientific History*. Paris, Scientifiques du Muséum National d'Histoire Naturelle. 767 p.

Woodroffe, C. D. 2002. *Coasts: Form, Process and Evolution*. Cambridge: Cambridge University Press. 623 p.

Zanchi, A. 1994. The opening of the Gulf of California near Loreto, Baja California, México: from basin and range extension to transtensional tectonics. *Journal of Structural Geology* 16, 1619–1639.

Ziveri, P., and Thunell, R. 2000. Cocolithophore export production in Guaymas Basin, Gulf of California: response to climate forcing. *Deep Sea Research II* 47 2073–2100.

Zwinger, A. 1983. *A Desert Country Near the Sea*. Tucson, University of Arizona Press. 399 p.

Markes E. Johnson holds the Charles L. MacMillan Chair in Geology in the Department of Geosciences at Williams College, Williamstown, Massachusetts, where he has taught since 1977. He earned a B.A. from the University of Iowa in 1971 and a Ph.D. from the University of Chicago in 1977. A fascination with ancient shorelines preserved as geological unconformities enticed Professor Johnson to first visit Baja California in 1989. He subsequently co-led annual research expeditions and field courses there involving students from Williams and the Universidad Autónoma de Baja California, Ensenada. Johnson is the author of *Discovering the Geology of Baja California: Six Hikes on the Southern Gulf Coast* (2002, University of Arizona Press) and numerous research articles on the geology and paleoecology of Cretaceous to Pleistocene settings in Baja California. He has also traveled in Australia, China, Siberia, Norway, and Canada to explore similar geological features that range from the Precambrian to Pleistocene in age. Johnson is a member of the Sociedad Geológica Peninsular and a fellow of the Geological Society of America, among other professional societies.

Jorge Ledesma-Vázquez is Associate Dean of the Facultad de Ciencias Marinas, Universidad Autónoma de Baja California (UABC), Ensenada. He earned a degree in engineering geology from the Instituto Politécnico Nacional (IPN) in Mexico City in 1977, a M.A. from San Diego State University in 1991, and a Ph.D. from UABC in 2000. He holds the national distinction of Investigador Nacional in Mexico, a rating associated with high performance by federal research agencies. His earliest visit to Baja California occurred in 1975 as part of a field trip with classmates from the IPN, for which he obtained funding through a personal appeal to the president of Mexico. He has co-led many research expeditions and field courses involving students from UABC, Williams College, and other institutions. With John Minch and sons, he is the author of *Caminos de Baja California: Geología y Biología Para su Viaje* (2003). Professor Ledesma-Vázquez is a founding member of the Sociedad Geológica Peninsular and has served as the organization's president since its inception in 1991.

Guillermo E. Avila-Serrano is a professor in the Área de Geología, Facultad de Ciencias Marinas, Universidad Autónoma de Baja California (UABC), Ensenada. He earned his B.A., M.A., and Ph.D. from UABC Ensenada in 1983, 1996, and 2005, respectively. He holds the national distinction of Investigador Nacional in Mexico, a rating associated with high performance by federal research agencies. His research as part of an international team is directed toward the study of taphonomy, and he has co-led many expeditions to the Colorado River delta involving students from UABC and other institutions. Professor Avila-Serrano has been a SCUBA diver since 1988, actively incorporating this mode of investigation with his research.

David H. Backus is a research scientist and part-time instructor in the Department of Geosciences at Williams College, Williamstown, Massachusetts. He earned a B.A. from Haverford College in 1982 and a M.A. and Ph.D. from the University of Washington, Seattle, in 1994 and 1998, respectively. A study of Cretaceous ammonites first brought Backus to the Baja California peninsula. Since 1999, he has been an annual visitor to the gulf as co-leader on field excursions sponsored by Williams College and Universidad Autónoma de Baja California. Backus has developed an expertise in remote sensing aimed primarily at the detection of carbonate deposits in the Gulf of California region. He has special interests in modern and ancient coastal dune deposits and the Mesozoic-Cenozoic biostratigraphy of peninsular strata.

Ann F. Budd is a professor in the Department of Geoscience, University of Iowa, Iowa City. She received her B.A. from Lawrence University in 1973 and her Ph.D. from Johns Hopkins University in 1978. She began research on the Miocene-Pliocene corals of southern California as part of her Ph.D. thesis and first came to Baja California in 2002 in collaboration with her former doctoral student Andrés López-Pérez. Budd's field of expertise is Cenozoic Caribbean corals and their changing biodiversity. Using a combination of molecular and morphological data, she studies the divergence between Atlantic and Pacific members of the families Faviidae and Mussidae over the past 50 million years.

Karl W. Flessa is Chairman and Professor in the Department of Geosciences, University of Arizona, Tucson. He earned a B.A. from Lafayette College in 1968 and a Ph.D. from Brown University in 1973. His primary research is focused on the environmental history and conservation biology of the Colorado River delta and the taphonomy and paleoecology of Recent and Pleistocene invertebrates in the Gulf of California. Together with Franz Fürsich,

Flessa is the author of *Ecology, Taphonomy, and Paleoecology of Recent and Pleistocene Molluscan Fauna of Bahia la Choya, northern Gulf of California* (1991, Zitteliana).

Matthew J. Forrest is currently pursuing a Ph.D. at the Scripps Institution of Oceanography, La Jolla, where he investigates ecological aspects of foraminifera relevant to their use as paleoindicators. He received a B.A. in English literature from the University of California, Los Angeles, in 1989, a B.A. in marine biology from the University of California, Santa Cruz, in 1997, and a M.Sc. from Moss Landing Marine Laboratories in 2004. While pursuing his master's degree, Forrest was introduced to the Gulf of California by Michael Foster and Diana Steller and subsequently became intrigued by shallow-water hydrothermal vents in Bahía Concepción.

Michael S. Foster is Professor Emeritus at Moss Landing Marine Laboratories, where he studied kelp forests, rocky shores, and rhodolith beds and taught related classes from 1972 to 2004. He earned a B.Sc. in physical science from Stanford University in 1964, a M.A. in education from Stanford in 1965, and a Ph.D. in biology from the University of California, Santa Barbara, in 1972. His first field research and encounter with rhodoliths in Baja California occurred in 1990, assisting his then master's student Diana Steller, who discovered the rhodolith beds around Isla El Requesón in Bahía Concepción. Foster is coauthor (with D. R. Schiel) of *The Ecology of Giant Kelp Forests in California: A Community Profile* (1985, U.S. Fish and Wildlife Service).

Oscar Gonzalez-Yajimovich is a professor in the Área de Geología, Facultad de Ciencias Marinas, Universidad Autónoma de Baja California (UABC), Ensenada. He earned a B.Sc. from UABC in 1979, a M.Sc. in marine ecology from the Centro de Investigación Científica y de Educación Superior de Ensenada in 1991, and a Ph.D. in geological sciences from the University of Southern California, Los Angeles, in 2004. He has traveled and explored Baja California since his first visit in the late 1960s and has conducted research in many of the region's coastal areas. His main research interests are high-resolution sediment records and their use as indicators of paleoclimatic and paleooceanographic change. He is an avid SCUBA diver and holds an Open-Water SCUBA Instructor rating. Gonzalez-Yajimovich is a founding member of the Sociedad Geológica Peninsular.

Ramón Andrés López-Pérez is a professor and researcher in the Instituto de Recursos at the Universidad del Mar in Puerto

Angel, Oaxaca, Mexico. He received his B.A. from the Universidad Autónoma de Baja California Sur (UABCS), La Paz, in 1996, his M.Sc. from Universidad Autónoma de Baja California, Ensenada, in 1998, and his Ph.D. from the University of Iowa in 2005. He began work on the Gulf of California's coral fauna when he first arrived in Baja California as a student at UABCS. He continues to study coral-reef ecology throughout western Mexico and has begun formal work on fossil corals from Chiapas in southern Mexico.

Astrid Y. Montiel-Boehringer is a Ph.D. candidate at Universidad Autónoma de Baja California (UABC), Ensenada, studying Pliocene-Pleistocene marine invertebrate communities in the central Gulf of California. She earned a B.A. in oceanography from the UABC in 1982 and a M.A. in marine ecology from Centro de Investigación Científica y de Educación Superior de Ensenada in 1992. She worked for the Environmental Protection Agency of the State of Baja California from 1993 to 2001, becoming Director of Environmental Analysis.

Héctor Reyes-Bonilla is a professor in the Departamento de Biología Marina at the Universidad Autónoma de Baja California Sur (UABCS), La Paz, where he has taught and conducted research since 1994. He received a B.Sc. from UABCS in 1990, a M.Sc. from the Centro de Investigación Científica y de Educación Superior de Ensenada (CICESE) in 1993, and a Ph.D. in marine biology and fisheries from the University of Miami's Rosenstiel School of Marine and Atmospheric Sciences in 2003. He has lived and worked on the Baja California peninsula for the last 25 years, absent only for doctoral studies. Together with several of his colleagues, Reyes-Bonilla is the lead author of the *Atlas de los Corales Pétreos (Anthozoa: Scleractinia) del Pacifico Mexicano* (2005, CICESE and a consortium of other organizations).

Rafael Riosmena-Rodríguez is a professor in the Programa de Investigación en Botánica Marina, Universidad Autónoma de Baja California Sur (UABCS), La Paz. He earned the B.Sc. at UABCS in 1991, a M.Sc. at San Jose State University in 1997, and a Ph.D. in botany at La Trobe University in 2002. His research interests are related to the taxonomy, ecology, and biogeography of marine communities dominated by plants. In particular, he is most interested in rhodolith beds and their significance as a key community around the world and has published numerous research articles and book chapters on this topic.

Eduardo Santamaría-del-Angel is a professor in the Facultad de Ciencias Marinas, Universidad Autónoma de Baja California (UABC), Ensenada. He earned a B.Sc. in oceanography in 1985, a M.Sc. in marine ecology in 1987, and a Ph.D. in 1994, all at UABC. His main research interests concern the application of satellite sensors to assessment of primary productivity in the ocean by means of false-color images, with a focus mostly on the Gulf of California. Since 1992 Santamaría-del-Angel has maintained the distinction of Investigador Nacional in Centro de Investigación Científica y de Educación Superior de Ensenada, a rating associated with high performance by federal research agencies in Mexico.

Diana L. Steller is a research biologist and the Diving Safety Officer at Moss Landing Marine Laboratories, where she has studied algal ecology and physiology, supported research diving, and taught field courses since 2001. She earned a B.A. from the University of California, Santa Barbara, in 1988, a M.Sc. from San Jose State University in 1993, and a Ph.D. from the University of California, Santa Cruz, in 2003. Her research activity in the Gulf of California began in 1990 with the discovery of rhodolith beds at El Requesón in Bahía Concepción. Since then, she has conducted many subsequent field trips, courses, and ongoing rhodolith investigations with colleagues in the Baja California Sur Ecological Society. She is a coauthor of *The Underwater Catalog: A Guide to Methods in Underwater Research* (2002, Shoals Marine Laboratory, Cornell University).

Miguel A. Téllez-Duarte is a professor in the Facultad de Ciencias Marinas, Universidad Autónoma de Baja California (UABC), Ensenada. He earned a B.Sc. from UABC in 1982 and a M.Sc. in oceanography and Ph.D. in earth sciences from Centro de Educación Científica y Educación Superior de Ensenada in 1987 and 2002, respectively. His research interests include mollusk taphonomy and paleoecology, stratigraphy, and more recently environmental geoarcheology and sustainable development. He has collaborated with Karl Flessa for the past 20 years on studies of the Colorado River delta regarding the human impact on natural environments recorded through ecological indicators. He is a member of the Sociedad Geológica Peninsular and honorary advisor to the Instituto Nacional de Antropología in Baja California.